Hector McLean

An Enquiry into the Nature and Causes of the Great Mortality among the Troops at St. Domingo

Hector McLean

An Enquiry into the Nature and Causes of the Great Mortality among the Troops at St. Domingo

ISBN/EAN: 9783337294465

Printed in Europe, USA, Canada, Australia, Japan

Cover: Foto ©berggeist007 / pixelio.de

More available books at **www.hansebooks.com**

AN ENQUIRY

INTO THE

NATURE, AND CAUSES

OF THE

GREAT MORTALITY AMONG THE TROOPS AT

ST. DOMINGO:

WITH

PRACTICAL REMARKS

ON THE

FEVER OF THAT ISLAND;

AND

DIRECTIONS,

FOR THE CONDUCT OF EUROPEANS ON THEIR
FIRST ARRIVAL IN WARM CLIMATES.

BY

HECTOR M'LEAN, M.D.

ASSISTANT INSPECTOR OF HOSPITALS FOR ST. DOMINGO.

Causa latet, vis est notissima.

LONDON:

Printed for T. CADELL, Jun. and W. DAVIES,
(Successors to Mr. CADELL) in the Strand.

M.DCC.XCVII.

TO

HIS ROYAL HIGHNESS

FREDERICK AUGUSTUS

DUKE OF YORK,

FIELD MARSHAL,

AND COMMANDER IN CHIEF OF ALL HIS MAJESTY'S FORCES,
&c. &c. &c. &c.

THIS ATTEMPT,

TO CONTRIBUTE, TOWARDS THE PRESERVATION OF THE
HEALTH OF THE BRITISH ARMY,

(Of which, in critical Times, HE has been chosen the Guardian;
and in the Conduct of which, HE has displayed,
not only, the Hereditary Valour of the HOUSE of BRUNSWICK,
but such consummate Prudence, and exact Propriety;
as at once, to merit the Applause of the PUBLIC, by His Vigilance,
and to command the Affection of the SOLDIER,
by His Attention and Kindness)

IS,

BY HIS ROYAL HIGHNESS'S PERMISSION

HUMBLY INSCRIBED,

AS A MARK OF THE MOST PROFOUND RESPECT,
AND SINCERE GRATITUDE—

BY

HIS ROYAL HIGHNESS'S

MOST OBLIGED AND
DEVOTED SERVANT,

THE AUTHOR.

St. James's Place,
26th June 1797.

PREFACE.

AFTER the publications, which have already appeared on the Yellow Fever; it may require some apology, for offering the present Work to the Public.

The Fever described by Dr. RUSH, which raged with so much violence at Philadelphia, differs very widely from the disease, which caused the astonishing mortality of St. Domingo. Nor does the pestilence of Boullam, as described by Dr. CHISHOLM, bear any striking resemblance to the Endemic Remittent, which made such havock among our troops.

They describe a fever highly pestilential, and contagious; whereas the fever of St. Domingo, never manifested any such disposition.

Having had the best opportunities, during a residence of nearly three years at Port-au-Prince; of observing the progress, and treatment of the St. Domingo Remittent; I thought it my duty to communicate the result of my experience, and the observations I made on the genius and type of the fever. It may again, be the fate of a British army, to visit this inhospitable island, and to suffer from its diseases; it is therefore of importance to collect every information, which may enable us, to avoid them, or to combat them with more success,

success, when they occur. Every practitioner ought to come forward, with his stock of facts, and observations; for the benefit of the Public.

The mortality at St. Domingo, has filled the minds of every one with terror and astonishment; and though men of distinguished abilities in their profession, assiduously attended the sick; their success was by no means, proportioned to their exertions or talents. When many minds, however, are occupied in one research, the subject is viewed in various lights; and, discoveries of importance may at length be made. Impressed with these sentiments, I have thrown together, the remarks and observations, which arose from an extensive experience; during a painful attendance, on the General Hospital at Port-au-Prince. To these I have added, whatever appeared to me connected with the welfare of an army, destined to act in a warm climate.

The constant occupation, my profession furnished me, did not permit me, to take down so many histories of the disease, as I wished; I was obliged to make my notes short; to retain only, leading and important circumstances. The points I have chiefly laboured to establish, are of considerable importance. I have endeavoured to prove, that what has been termed the Yellow Fever of St. Domingo, is is not an infectious difeafe; that it is not a new or peculiar diftemper; but the common Remittent Endemic of that country, applied to the Englifh conftitution, and accompanied occafionally with yellownefs, as an accidental fymptom. The dread of its being infectious, has injured the recruiting fervice, by terrifying young men from enlifting in any Weft India regiment; and many have been kept in a ftate of continual alarm and terror, when the fervice required them to have the leaft communication with the fick. It is pleafing to reflect, that the general teftimony of all the phyfi- cians at St. Domingo, declares that the Remittent of that ifland is not contagious. One fource of fear, is thus removed; a fource, which has

unne-

unnecessarily, alarmed and terrified all those, who embarked for this climate.

I have endeavoured to show the causes, which render the Western climate peculiarly dangerous to our youth; and I have recommended a scheme of recruiting men, for this service, at a more advanced period of life. The chance of living, in a warm climate increases, as we advance from thirty-five to fifty years of age. Men at these periods, may enjoy health at St. Domingo, and perform active duties.

I have founded my plan of Prevention, on the theory I adopted respecting the causes, which rendered the Remittent so destructive.
The

PREFACE.

The rules for diet and exercise are the result of experience, and observation, and I trust will be found of real utility; as they apply to all warm climates.

The method of Cure, which I pursued, after many fluctuations of opinion and practice, may be thought too bold, by those who have not seen the rapid progress of the disease. It however happily succeeded, in many instances; and I have candidly stated the grounds on which it was founded, and its various success. In a disease, which baffles ordinary means, the physician must seek, extraordinary resources, and endeavour to subdue by vigour, what would not yield to common expedients.

It

PREFACE.

It is probable, that in the courſe of the work, I have made obſervations, which have been formerly made by others on ſimilar ſubjects.—It is not eaſy to avoid this, in medical reaſoning; if it occurs, it ariſes from my ignorance of the authors, for I had not the power of peruſing any books during my reſidence in St. Domingo. It was a duty, I owed to the Army with which I ſerved; to publiſh for their benefit, whatever might enable them to avoid danger, or ſecure their health. To that Army, I owe many obligations, which I ſhall always acknowledge, and remember, with gratitude. If this work contributes to ſave one life, or to introduce a better mode of treating this formidable diſeaſe; I ſhall feel

myſelf

myself more than rewarded, for my labour.

It is a pity, that Officers in command, do not read such parts of medical works, as treat of the health of soldiers. This kind of knowledge, in warm and unhealthy climates, would prove highly useful. The General of an army, ought to be well informed, in whatever regards, the Encampment, Diet, or Exercise of his soldiers. It is not in the power of medical men, to pursue extensive plans, without the support of the Commanding Officer; they can only recommend, but cannot execute. Their schemes of health, are more readily complied with, when the officer understands fully, the principles on which they

are

are recommended. The works of Jackson, Pringle, and Monro, contain valuable information on this subject.

Whilst I venture thus before the Public, I must claim their indulgence; I am fully aware of the imperfection of my essay. It was written in the bustle of a few weeks leave of absence, from the duties of my station; to which I am again speedily to return. The language must often offend the ear; and the arrangement is not so correct as might be wished. I have endeavoured to state, in the clearest manner I could, whatever I thought interesting or useful. I have blended the matter of fact, and my reasoning, too much together; but my time

time would not permit me to alter this arrangement, which would oblige me to new-model the work. These, and other suggestions, were made by a friend, celebrated for his talents and learning; who condescended to peruse the work, and whose remarks, would insure it a better reception, if I could avail myself of his plan. I trust this will form some apology, for the general incorrectness of the performance.

I shall conclude, by observing, that many practitioners have lost their patients, by searching for definite indications, in the fever of St. Domingo. Time has been lost, in combating the lesser effects, or symptoms of the disease. My object has been, to alter, if possible, by

sudden

sudden and powerful means, all the circumstances of the habit, and by this change, to give rise to a new train of movements. It ought to be ever remembered, that when we overcome any morbid action, in the human body, or suspend its power; that the system is immediately disposed, to return to its usual laws. So that to suspend the action, of a morbid power, in the body, is to afford the system a cessation from hostilities; if I may so speak, and to enable it to recover its ancient constitution.

Upon the whole, I have endeavoured to make this Enquiry as useful as possible to the Army, for whom alone it was written.

CONTENTS.

INTRODUCTION - - Page 1

CHAPTER I.
SECT. I.

Character of ST. DOMINGO - - 7
Situation of its Towns - - - 8
Causes of Ill Health - - - 9
State of the British *in Port-au-Prince* - 12
Difference between the French *and* English *Constitutions* - - - 14
French *Medical Practice examined* - 18

SECT. II.

Further Considerations on the Causes of Ill Health - - - - - 23
Miasmata considered, and the Effects of Moisture - - - - 24

	Page
The Effects of Debility on the Vascular System—Determinations arise from Debility	27
Chemistry the great Source of Improvement in Medicine	31
Yellowness, not always caused by Bile	32

SECT. III.

Plethoric Habit, dangerous in the Remittent	36
Young Men more exposed to danger than others—Women and old Men escape the Fever	37
Practical inferences, from these Facts	38
Scheme of recruiting Men at a certain Period of Life	39

SECT. IV.

Causes which retard Medical Improvement	41
Surgery and Medicine compared	43
Of Proximate Causes in Disease	48
Our Ignorance of these prevents Improvement	52

CHAPTER II.

SECT. I.

	Page
The Difficulties of prescribing by Indication	56
Absurd Reasoning on supposed Proximate Causes	58
Morbid action constitutes Disease	63

SECT. II.

Opinion of the Yellow Fever—It appears to be the common Remittent	71
It is not the Fever of Philadelphia or Boulam	73
The Yellow Fever not contagious	75
Grounds for this Reasoning	78

SECT. III.

Causes of the Frequency of the Remittent	81
Similar Causes produce Remittents and Intermittents	85

CONTENTS.

	Page
State of the Organs in the Remittent	86
Phænomena in the Remittent	91
Critical Days	93
Causes which excite the Fever	94
Diagnostic difficult	96
Prognostic	97

SECT. IV.

Two Cases of the Remittent—French Treatment of them	104
Dispute in Jamaica on the Nature of the Fever	111
Practice of the Gentlemen at St. Domingo	115
Various Remedies examined	119
Calomel; Blood-letting; &c.	121
Cases detailed	131
The Author's Treatment of the Remittent after considerable Experience—Comparative Success of this Method	164
Practical Remarks	172

CHAPTER III.

SECT. I.

Means of Prevention—Changes in the System from Heat—Preparative Course for a Hot Climate - - - 192

SECT. II.

Method of treating Troops after landing—Situations to be chosen for their Residence - - - - 202
Manner of Exercise recommended - - 208
Different Posts examined - - 215

SECT. III.

Considerations on General and Regimental Hospitals - - - 229
Hospital Corps, unfit for their Occupation 235
A Medical Board, recommended with large Armies - - - - 241
The Qualifications of Physicians examined 243

SECT. IV.

	Page
Considerations on Diet adapted to the Climate	250
Claret and Madeira compared	252
Diet recommended by the Author	255
Smoking considered—Its Effects	262
Flannel examined	268
Cotton Shirts recommended	269
Bilious Complaints	271
Causes of Bile	279
Cure	281
Of the Prickly Heat—Its Causes, and Nature	293
It is not dangerous, or critical	299
Bathing not dangerous, when it is out	300

APPENDIX.

APPENDIX.

	Page
Reasoning of the Author confirmed by Practice	309
In Intermittents	310
Fevers	318
Ulcers	328
Lues Venerea	336
Small-Pox	341
NOTES	347

ERRATA.

Pp. 310 to 318. the top lines, *for* The Author's Practice in Intermittents, &c. *read* Appendix.

INTRODUCTION.

THE present Enquiry proposes to investigate the nature and causes of the astonishing mortality among the troops in St. Domingo; a mortality almost unequalled in the annals of war, and which has nearly annihilated our army in that quarter, or rendered them incapable of energy and exertion. In this view, it is an enquiry of the utmost importance, as the preservation of so many valuable subjects is involved in it. The opportunities I have had of observation have been numerous and extensive: a residence of nearly three years at Port-au-Prince, which might be considered rather as a general hospital than a garrison, enabled me to mark the progress of that formidable disease, which in this country has obtained, without distinction, the name of Yellow Fever. At a very early period, I could not help remarking the unequal warfare carried on, in that quarter, between an European army and a people inured to the climate. The European soldier, languid and relaxed, from the excessive heat, had to contend with an enemy, inferior indeed in the art of war, but formidable from a frame of body which was adapted to the climate, and derived vigour and activity from that influence of the sun, by which foreign troops were enervated and exhausted. In a warm climate, the European soldier requires many comforts; but the naked Brigand climbs the tree for his daily food, and sleeps in perfect security under the broad canopy of the sky. A contest with such a people must be ever tedious, unequal, and uncertain. We cannot pursue them to their fastnesses. The neighbouring mountain affords them the same protection and means of existence with the one from which they have been just driven, and an extensive fertile country affords them perpetual change. Such a people can only be brought into subjection by a treaty, or overcome by an army of negroes, possessed of the same habits as themselves, but more expert in arms, and led on by such a proportion of European troops as might animate and encourage them. The armies of India are organized on this principle; and I am convinced the armies of the West would become more victorious by similar management.

I believe

INTRODUCTION.

I believe Sir ADAM WILLIAMSON had in contemplation the commencement, and actually begun this scheme of war, which his return to England prevented him from compleating. Under an officer so deservedly popular, and to whom the natives were so strongly attached, such a plan would have produced the most beneficial effects; the lives of thousands, who have fallen, not by the sword of the enemy, but by the climate, would have been spared; and the conquest of the island would become more certain and more rapid. The temporary expence of the establishment would no doubt be increased; but have they diminished under an opposite plan? Do we not incur an enormous expence in the hire of transports, and the attendance of ships of war to convoy them? and do we not suffer still a greater loss in the destruction of so many of our valuable troops?

So much I have thought it my duty to say on the general plan of our operations. Some Europeans we must have there, but their number ought to be diminished, and they should be formed from the veteran remains of the regiments who have already served there, and overcome the terrors of the climate. To such men an additional pay might be granted; no reward can equal the hazard of the service. I am doubtful whether it would be found policy to accept the submission of the Brigands, were they all inclined to submit at the present moment; they are too numerous to be trusted; and should they once more bend under the lash of the planter, their habits of indolence and depredation would soon return, and they would again sigh for licentious idleness. Among new negroes they would be so many apostles of sedition, and they would scatter amongst them the creed of the national convention; a creed which would make revolt and murder duty, and which would dignify every act of horror with the sacred name of an enthusiasm for freedom. The only scheme of subduing them appears to me to be, the enlisting great numbers of them into our army, and forming them into regiments commanded by British officers, or French loyalists of approved fidelity. Military discipline, without extreme rigour, would beget habits of subordination and attachment to their officers; and marks of distinction, judiciously bestowed amongst them, would generate military pride, and an ardour very necessary to connect and support such bodies. The rest, whom we could not employ in this manner, must be protected in some district, as free men,

under

under the auspices of our government and the guardianship of our laws, or they must be sent to some other country, or divided and watched among the several estates now to be re-peopled: to exterminate them entirely is perhaps impossible, and were it possible, would not, I trust, be recommended by Great Britain, who would not renew the scenes which stain the Spanish annals; scenes which deluged the peaceful plains of Hispaniola with the blood of that meek and patient race, who were its natural possessors.

These observations, though not strictly medical, are by no means foreign to my purpose. My object is to diminish the mortality of British soldiers in St. Domingo; and such a plan would more effectually answer the purpose, than all the medical exertions of the most experienced and skilful physicians. The errors committed in the choice of situations for the army, on our first taking possession at St. Domingo, have been severely felt ever since. Misled by erroneous and interested advice, they were conducted to Port-au-Prince, the most unhealthy spot in the island, where they languished and dwindled away without any service to the cause they were meant to support.

support. Even the advantages gained by the enterprize of the gallant MARKHAM were hardly any compensation for his own individual loss, and the many others who perished in supporting his exertions. A few frigates cruising in the Bite of Lugan would have gained more advantages than any force we could station at Port-au-Prince, which has proved the grave of our Army, and which must have ultimately fallen into our possession.

I shall now proceed to the immediate business of this work; to enquire into the causes of the mortality that has distressed our forces.

CHAPTER I.

SECT. I.

Character of ST. DOMINGO—*Situation of its Towns*—*Causes of Ill Health*—*State of the British in Port-au-Prince*—*Difference between the* French *and* English *Constitutions*—French *Medical Practice examined.*

ST. DOMINGO, from the earliest accounts of its settlement to the present period, has been an island remarkably unhealthy. The Spanish records bear the most unequivocal testimony of the rapid and destructive progress of its diseases, which unpeopled their rising villages, and disappointed their precipitate rapacity in the pursuit of many of its favourite objects. Since the French have had a share of this valuable island, they have also experienced the fatal effects of its climate. It was their policy to conceal the ravages of disease, and to induce as many as possible to colonize and settle in this most luxuriant and fertile country. They succeeded; and many daring adventurers took up their abode in this fruitful region, from which they seldom entertained any wish to return to the mother country. Even noble families obtained grants of lands, and sent their spurious descendants to occupy them, who, in this new habitation, enjoyed every luxury, and the mimic splendor of the noblesse themselves. These adventurers never wished to return; and they accordingly cultivated and adorned their plantations, as the residence of their youth and age. This, joined to the fertility of the country, is one reason why the island of St. Domingo was more highly cultivated than any of ours. The planters and colonists of our English islands seldom pass their lives abroad, and look with fond expectation to the day they are again to revisit their native soil; hence their plantations are not so highly finished in the culture, nor their establishments so splendid or permanent as those of the French. The English planter consults present advantage; the French, looks further forward.

The towns of St. Domingo, especially Port-au-Prince, are admirably calculated for

for the purposes of commerce. Considerations of health gave way to the schemes of avarice and the convenience of attaining riches. It must be confessed, that the French made the utmost of every situation they occupied, and improved them as far as they were capable of improvement; their streets were wide, extensive, and open; a chain of virandas, or piazzas, sheltered from the sun, connected all the houses, under which the passenger could walk free from every inconvenience: a row of trees on either side of the street refreshed the eye, and gave a rural appearance to the whole, whilst streams of water, flowing along, carried off any impurities, which, in spite of every care, might accumulate. The French have taken great care to supply all their towns amply with water; a great consideration in hot countries, but which we neglect in our colonies in a remarkable manner. But notwithstanding these endeavours on the part of the French, they occupied situations, which could not be rendered healthy by any means they employed. Port-au-Prince is one of those. It is placed at the bottom of an immense bite, which pushes itself into the heart of St. Domingo. The scite of the lower part of the town is, in fact, on a marsh gained from the

sea,

sea, the skirts of which are covered with weeds or mangroves, where decomposed animal and vegetable matters are promiscuously thrown; on these the sun exerts its power, and the breeze conveys the noxious particles with a new activity to the lungs and bosoms of the inhabitants. But this is not all: the sea breeze, which in other situations is hailed as the genial source of refreshment and health, is here interrupted; the island of Gonàve is so placed in the mouth of the harbour, as, in a great measure, to intercept this salutary gale; and, before it arrives at Port-au-Prince, it loses its usual coolness, by passing over heated lands, and gathering in its course noxious vapours. This necessarily results from the inland situation of the town. Besides these manifest causes of ill health, Port-au-Prince is exposed to the action of others. It is placed on a level, on the verge of the bite, and surrounded by very lofty mountains, from the bottom of which a horizontal plain stretches towards the town. Torrents of water, in times of rain, rush through this plain, and retain their impetuosity till they reach the sea.

The land is moistened, but after the torrent ceases the water stagnates; small streams,

attaining

attaining a horizontal level, lose the impetus acquired in their descent; they linger in the plain, and by mingling with the soil form a marsh. On this marsh a vigorous sun acts daily, and evaporates its noxious particles, which are conveyed to the lungs of every one that breathes, and applied to their skins, and probably in this manner communicate with the blood. This is a never ceasing cause of disease, a nursery constantly rearing mortal poison. In every inspiration, we draw into our bosom a column of air thus impregnated, in every step we walk, a fresh application of these particles is made to our bodies; it is no wonder then, that on this fatal spot the British troops caught fever in each treacherous breeze. It is true, that the French, when they exclusively possessed this town, did not perish in the same proportion with us. The causes of this difference are not difficult to trace; the French possessed a free open country, and could at pleasure retire to breathe the more pure atmosphere of their distant plantations. Every merchant, every planter, in short, every inhabitant, possessed the power of retiring into the country and changing their situation.

In this manner they obviated the fatal effects which would otherwise result from the uniform and constant application of the exhaled miasmata. They were besides amply supplied with fresh vegetables, and every luxury that contributed to pleasure or health. Very different was the situation of the British; surrounded on every side by the enemy, they were imprisoned within the walls of a town half demolished, daily exposed to the sources of disease, and without a supply of good vegetable or animal food. Instead of the cheerful elevation of spirits, which the view of prosperity and peace naturally produces, the British were depressed by every thing that could sink the mind to a state of despondency. In want of the comforts that can render war or exile tolerable, and exposed to an unfriendly sun, they became the daily spectators of death. The constant ravages of fever amongst them spread a general gloom, and weakened the vital powers; a moment of debility is favourable to the invasion and the conquests of disease, and accordingly thousands perished; besides, the British, from the numbers who daily expired, were so weakened, that a large portion of duty was thrown on those who were well, or even convalescent.

Thus

Thus difeafe, by debilitating and expofing the few who had efcaped direct attacks, neceffarily perpetuated and increafed itfelf. The French inhabitants were chiefly compofed of people born in St. Domingo, and fupported by that gradual adaptation to the climate, which habit confers on the natives of the moft unhealthy regions. All the Englifh inhabitants were chiefly ftrangers from northern climates, and little habituated to any warm region. A ftate of war too expofes all claffes, in a befieged or blockaded town, to numerous difadvantages, but chiefly the foldier, who muft brave every viciffitude of climate, the frequent changes of each varying night, without the benefit often of refrefhing reft or reftorative diet. In peace it is otherwife; there is no caufe for the fame vigilance and perfeverance; the foldier carries on his duty with all the comforts of civil life. Port-au-Prince had all the difadvantages of a blockaded town; the foldiers or inhabitants could not change their fituation, or fly to the hills from the ravages of difeafe. Danger furrounded them in every quarter, and they were obliged tacitly to fubmit to their fate.

The French poffeffed other advantages.
Their

Their conftitutions feem better calculated for warm climates than ours. The manner of life in old France, or its colonies, was not calculated to form that irritable, fanguine, plethoric habit, in which the effects of inflammation are particularly felt, and in which they proceed with rapidity to a fatal termination. The French, ere they vifited thefe colonies, either for a tranfient or permanent refidence, were in a ftate of falutary preparation; they took feveral medicines calculated to diminifh the plethoric ftate; they continued, on their arrival, to purfue a moderate, cautious, and fober plan of life; the quantity of animal food they ufed was very fmall; they indulged not in wine, or ardent fpirits, whilft the body was kept moderately open, and they were gradually accuftomed to bear the fun; the tepid bath too was ufed to cleanfe the fkin from impurities, and preferve it open, for the purpofes of perfpiration, fo effential to health. The Englifh, on the contrary, who embarked for war, were in every refpect the reverfe of the French. Their diet, compofed of large portions of animal food, and amply diluted with fermented liquors, rendered them full and irritable. In this ftate they embarked; in this

ftate

state they landed; superadding the effects of salt provisions at sea. Instead of the preparative and sober regimen of the French, we are inclined to prolong our convivial enjoyments, and sacrifice considerations of health to gaiety and friendship. Some are impressed with fear, and imagine that debauches of wine banish not only their fears but their danger. This is an opinion fatal to many. They land, and, friend meeting friend, rush to the feast, where, to use the words of ADDISON, " death and disease lie in ambuscade among the dishes." On the early arrival of the British too, after the long confinement of a passage, they run about the streets, careless whither, and expose themselves at once to the sun's most powerful influence. Ere habit has imparted its defensive powers, they are ordered to duties which require activity, and which increase the irritability of the constitution, and, in this situation, as it were betrayed into debility, they are seized with fever, which finds them an unresisting prey.

In this manner that astonishing havock, which has terrified and alarmed abroad, and excited just fears at home, was created, and continued. If I mistake not, it will be found,

on

on a strict examination, that a tendency to an inflammatory habit forms the basis on which the Remittent of St. Domingo establishes its devastation. The sudden determinations made to the vital organs depend entirely on the energy and irritability of the vascular system. In the French constitution, such derangements do not readily take place, because there is not in the vascular system either the same energy or irritability as in the English.

The re-action, as it is termed, is weak and feeble in the former; in the latter, is strong and powerful; hence fewer of the French perish than of the English, when actually attacked by fever. But this difference in no degree depends on the superiority of their medical practice, which they arrogantly assume. It is with pain I am obliged to mention, that I have not met with one French practitioner, on whose veracity the least reliance could be placed. Impelled by vanity, and that spirit of gasconade in narrative, which of late fills all their writings, they exaggerate every thing they relate, and, with deliberate impudence, attribute to themselves the most extraordinary talents. Such a charge against a set of men, professing at least a liberal pursuit, ought to be be very firmly and well supported. I appeal, without hesitation, to every medical gentleman who has practised with them in the West Indies, and I do not fear the least contradiction. Equally œconomic of medicines and truth, they committed their patients to a nurse, and left the issue to nature. If there is any thing in the powers of medicine, or in vigorous treatment and prescription, it is wholly wanting in the French practice. Ptisans and frequent injections form the prominent features of their treatment. The German dieta aquæa, and large quantities of lemonade, are also prescribed. Camphor, opium, musk, and æther, are very rarely given; except in the hopeless stages of fever, and then not in quantities to produce any great effect. I do not deny that the French physicians merit praise for their attention to a very essential part of medical treatment. Nursing is often of as great importance as medicine, and they have improved it; but they possess no higher merit. Their plan forms a very excellent part of a system of practice, but cannot be surely trusted wholly. Let the French and English modes of treatment be blended and mixed; the vigour of our practice, with the benefit of French nursing, and the patient will have every chance the

present

present state of medical knowledge can afford him.

On my first arrival at Port-au-Prince, a French physician of some learning and abilities, almost induced me to believe, that he had a mode of treating the Endemic of that country, which generally succeeded. Trusting to his conversation, and confiding in his skill, I took him to visit one of my assistants then under fever; I requested he would prescribe; he took the management into his hands, and the youth perished. This would be a solitary instance of no great weight, as it might happen with the ablest physician, were it not followed by a number of others equally notorious. But the physicians always had some silly subterfuge, such as not being called in time, the obstinacy of the patient, and a variety of other trifling evasions. But I shall not rest the proof of what I have asserted on this alone. Dr. JACKSON, whose abilities and learning are universally known, and whose liberality and candour have been experienced by all those who have ever met him in his professional character; consigned to the care of a French physician, a certain number of the German and Dutch troops. From his pompous and confident assertions, much was expected; it was a fair experiment, and a liberal comparison between French and English treatment. What was the result? On a comparison of the returns from that hospital, the number of deaths considerably exceeded our loss with similar numbers. This was an unequivocal trial, and amply refuted all their exaggeration. The truth was fairly stated, and made them silent.

I would not be thus particular in examining the proceedings of the French physicians, had not they, in the most illiberal and shameful manner, propagated unfounded reports relative to English practice. They boldly asserted, that the English physicians and surgeons, killed their patients, and were ignorant of their profession. Such a report, industriously circulated, had bad effects; it diminished the confidence of the troops in the medical staff; and confidence is of the utmost consequence, as it supports the mind, and elevates it with hope. They were not contented with scattering these rumours at St. Domingo, but they also wrote home, and addressed an anonymous letter to the medical board, and even put paragraphs in our papers, all tending to lessen the confidence of the British in the medical officers who attended them. I trust I shall be excused for entering into this detail, and stating the question fairly. Every one will allow that there is something to be learned in the diseases of every country by being on the spot, which neither description or reading can supply; nor can it be supposed, where human lives are at stake, that any one would for a moment hesitate to adopt the plan of others, where it promised any hope of success. The man who from pride, prejudice, or illiberal rivalship; would reject improvement, or despise knowledge, ought to be expelled society as a monster, who would sacrifice the human race to his unfeeling vanity; but in no place was improvement more likely to be adopted, than St. Domingo, where, unfortunately, all our plans of treatment had but little success.

The medical staff of that island was composed of men of liberal education, and to whose attention the army will bear testimony; it was not likely that such men would sacrifice, to an illiberal prejudice, the lives of their countrymen, endeared to them by acquaintance, and by common hardships.

It might be supposed, and in fact it was given out, that St. Domingo, before our arrival there, was by no means unhealthy; and that

that their garrifons fuffered very little. In this declaration the French phyficians exulted; but they attributed the difference entirely to their own fkill and fuperior management. We have already feen how much credit is due to their affertions; and in this, as well as in moft inftances, they will be found falfe. Every body knows, that St. Domingo was always unhealthy. The moft falubrious and beft climates have their periods of difeafe; there is no country yet known that can boaft an exemption from thefe laws of nature. It has been already ftated, that the French fuffered lefs than we have done; but the fame fever, which has fwept off our troops, raged alfo in the French garrifons, and made great havock. There cannot be a ftronger proof given of the mortality of the French troops, during the old government, than their being obliged, every three or four years, to renew the garrifons with 10,000 men. What became of thefe troops? few of them ever returned to France. They became victims to the fever of the ifland. When the revolution begun its horrors, it was not in the power of that convulfed nation to fend the ufual fupply to St. Domingo; confequently the former garrifons were greatly diminifhed;

and, in fact, we found very few of the veteran troops of Old France to oppose our views or progress. I have been thus full and explicit, on the former state of this island, to disprove the false assertions of the French physicians, and refute insinuations equally unfounded and illiberal; but which, by creating despondence and fears, and lessening confidence, might produce the worst effects.

Hispaniola has always been, and I fear in a certain degree will always continue, unhealthy; but our further progress into the country, by enabling us to change our situation, by varying the scene, by amusing and delighting the senses, and being enabled to procure a more wholesome diet, may yet dispel our fears, and inspire more confidence. Such a change would gratify the feelings of humanity, and be particularly pleasing to those who have witnessed the distressing scenes of our early operations in that island. I shall proceed in the next section to examine more particularly the mode in which the fever is produced, and the manner of its operation on the human body.

SECT. II.

Further Considerations on the Causes of Ill Heath—Miasmata considered, and the Effects of Moisture—The Effects of Debility on the Vascular System—Determinations arise from Debility—Chemistry the great Source of Improvement in Medicine—Yellowness not always caused by Bile.

IN enumerating the general causes which produce, or rather dispose, to ill health, at St. Domingo, I omitted to mention the Land Winds, which prevail more particularly in June and July, but are more or less felt throughout the year. These breezes blow in the morning, and have a pleasing coolness at an early hour; but they sometimes continue for days, and even weeks, and as the day advances become hot, drying, and unpleasant, resembling the *Siroc* of Naples. These winds suddenly check perspiration, the great source of coolness and health; the skin becomes parched and dry, and there is usually a sense of oppression in breathing, and a tendency to sigh. These winds very generally excite fever, and bring on paroxysms anew, in convalescents. The French shut their doors and windows against them, and go out as little as possible whilst they prevail; but their effects are sensibly felt.

Let us now examine the causes, which more remotely seem to produce the fatal fever in question. It would seem that it arises from elastic fluids, miasmata, or vapours, applied to the human body, either through the medium of the lungs or skin, or perhaps swallowed with the saliva. This opinion is rendered probable, by observing that the fever arises more frequently, and proceeds with more severity, in situations where a brisk evaporation from stagnant marshes is going forward. It would seem that rain, simply as such, or dew, do not produce any derangement in the human system, except what may arise from simple obstruction and the effects of cold. I have myself been exposed to the dews, without the means of shifting, for two nights and upwards, when a prisoner with the Brigands, without feeling any inconvenience. Doctors JACKSON and

and MASTER were similarly situated, without any bad consequence; and in coming from Jamaica to St. Domingo I have slept several nights on deck, with a very slight covering; I never found any illness result, though I have been wet for several hours. The longer, water has been incorporated with the soil, without fresh rain, the more virulent and dangerous the miasmata become. Thus, the seasons in which rains prevail are found more healthy than when dry weather has continued for any time: in the first case, the rain is again evaporated before the soil has imparted its pernicious qualities, but in dry weather it has remained, so as to acquire the fatal activity it exerts on the system. In what manner the action of the sun, and the admixture of a certain soil, produces this activity, can perhaps, scarce be explained in the present state of chemical knowledge. It must be admitted too, that fatal miasmata arise where there are no very certain appearances of a marshy soil. The Mole and St. Mark's, do not appear surrounded with marshes, yet the fever reigns in both these places with great activity. Miasmata may certainly be wafted with the land breezes in some activity to a considerable distance; they

are

are undoubtedly more powerful the nearer they are to their source, and in time become changed by admixture with the atmosphere and distance from their origin. Thus ships, lying in harbour at a certain distance from the shore, are liable to the fever of St. Domingo, but if they go to sea, or cruize at a greater distance from the land, the crew become more healthy, and less obnoxious to disease. It is not to be doubted but the sun acting on moisture, almost in any soil, will produce disease. The grand defect in our knowledge, arises from our being ignorant of the particular nature, of the elastic fluid or miasma; which arises from marsh. If this point was once ascertained, we could reason on the varieties produced by a difference in the soil, and the less or greater action of the sun; so as perhaps to throw some light on varieties in disease, which seem to arise from these causes. How these miasmata or vapours produce their effects on the body is not distinctly known. Let us attend to the phænomena which they produce, and try to account for them on principles already ascertained, or rendered highly probable, from the phænomena themselves.

The firſt evident effects are, debility and languor in many of the important functions of life. The vaſcular ſyſtem, in ſome of its ſubordinate diviſions, appears greatly weakened; for although there ſeems a degree of excitement and action, yet irregular determinations of blood take place to various important organs, ſuch as the head, ſtomach, liver, and lungs.—Determinations of blood cannot happen to any particular organ without debility in ſome part of the vaſcular ſyſtem, which deſtroys the balance eſtabliſhed for a juſt circulation between the propelling power of the heart and the reſiſtance of the arteries. In ordinary caſes, where the propelling power of the heart is not altered, if there is any weakneſs in a particular ſet of veſſels, a determination happens; but in caſes of fever, whilſt the veſſels ſeem to loſe greatly their reſiſting power, the energy of the heart is remarkably increaſed. From this cauſe the danger of determinations during fever is greatly augmented. In what direct manner all this is produced eludes at preſent our keeneſt reſearch. It would ſeem, at times, that the miaſmata attacked at once the very principle of life; from the beginning, in ſuch caſes, all energy is ſubdued, and the ſufferer gradually periſhes
under

under a fenfe of oppreffion only, and a flow diminution of animal powers.

Many inftances of this kind occurred, which at firft flattered the unexperienced fpectator, but which ftruck the attentive obferver with impreffions of the greateft danger. Where there is a fenfe of pain and a re-action, the ftock of fenfibility and vital energy, is yet entire, and may be acted on with fome hope; but where thefe are greatly impaired or fubdued, the caufe of the difeafe is powerful, and will in general be victorious. In proportion to the utility and neceffity, of any organ to the maintenance of life, will be the danger of a determination to it; thus the brain, the lungs, and the ftomach, as being highly neceffary to fupport the living phænomena, are the moft dangerous feats for thefe derangements. In the brain, either its direct functions are deftroyed by fubtile caufes, or by more evident preffure from effufion on it, or by the deftruction of its organization. The ftructure of the lungs is frequently deftroyed by the fudden progrefs of inflammation, which, by diminifhing their capacity, and the eafe with which their effential functions are performed, has a direct tendency to deftroy life, which we feem seem to renovate and strengthen from this wonderful organ. But besides the direct influence the lungs have on the vital principle, any impediment in their functions creates new determinations in the blood, and extinguishes life by the slow progress of partial disease.

The stomach, the most important support of all our functions, becomes, in the fever of St. Domingo, a principal seat of determination. Very early in the disease, an immense quantity of blood fills its vessels, inflames its inner coats, and begets in it a sensibility and irritability almost incredible. In vain it is attempted to throw in any medicine, in the the most soothing or lenient forms; the most grateful cordials, and the most insipid liquids are thrown up with a celerity equal to an explosion. The effort to reject is made before any thing almost touches the stomach. These efforts to reach are continued often till death. From the great determination of blood, and the violence of the action in the vessels, immense secretions are made in the stomach, which, acquiring there a dark colour from the admixture of other fluids, and perhaps portions of the coats of the stomach, is
called

called the Black Vomiting, generally a very fatal prefage of the event; at length the coats themfelves are feparated and detached, and a mortification, affording a temporary fufpenfe from agony, clofes the cruel fcene.

During this progrefs there is little remiffion; but in general a flight abatement of the fymptoms occurs towards noon, and an exacerbation foon follows. What fecret law of the animal œconomy influences and determines thefe periods and fluctuations of fever are not at prefent, nor perhaps ever will be known. In fome inftances, where one paroxyfm has finifhed the difeafe for a given time, the whole would appear to refemble a chemical procefs, where a certain time was required to complete it; but the fubfequent renewals, and the fimilarity of the fucceeding exacerbation, obliges us to abandon fpeculative opinion, and confefs our ignorance. In violent and rapid cafes, where a fudden recovery has happened, it has fometimes been preceded by a tormenting intolerable pain in fome particular finger or toe; at laft, a livid fpot, with an extended rofy bafe, makes its appearance, and the difeafe is gone; as if a certain chemical combination had happened,
and

and the refult had efcaped by a filent but vigorous effort, like the electric fluid.

On the difcovery of the real nature of the marfh miafma, of the various circumftances which affect its condition, on which its ftrength and noxious powers depend, is founded the future improvement of this intricate part of medicine. That new chemical combinations are formed, deftructive to the principle of life, no one will deny, who has ferioufly thought on the fubject; in no other way can the fudden derangements be in any manner accounted for. The mechanic effects of miafmata would, from their very nature, be imperceptible, and, though we are utterly ignorant of the direct chemical changes or proceffes, yet it may be inferred, with more probability than any other fuppofition, that they really exift. Chemiftry is now unfolding more intimately the nature of the animal fibre; and if human abilities, aided by fcience, promife any light on fubjects which nature has involved in darknefs, we may entertain fome hope of improvement. —But chemiftry mourns the death of LAVOISIER, of him, whofe labours made it a new fcience; and promifed to fcatter light over the

the darkeſt inveſtigations. The loſs of a man, who devoted his time and fortune to the moſt enlightened purſuits, with ſo much happy ſucceſs, and perſeverance, will be felt by future generations, and will deſcend to poſterity as one of the great ſtains of the revolution.

Many appearances induce us to believe, that very conſiderable changes go on in the fluids. The ſudden appearance of livid ſpots, the oozing of blood from all the mouth, and its rupture from the noſe, cannot be accounted for from the ſolids only; both are uſually affected, and muſt be ſo from their very intimate connection, and ſtanding in the relation of affecting each other, as cauſes and effects of many phænomena.

The yellowneſs, which is certainly no favourable appearance, and gives a peculiar name to the fever, does not always ſeem to depend on the mixture of bile with the blood. It is not eaſy to ſupport the opinion I am about to offer, becauſe the facts which would render it clear are not eaſily obtained. The ſame opinion was formed by my friend, Dr. MASTER, before we had ever converſed on on the subject, and on the same grounds. The yellowness appeared to us, to arise from a change effected in the colour of the serum, dependent on a peculiar action in the vessels; because jaundice, or a mixture of bile with the blood, is not in itself a disease suddenly fatal, or even very formidable; and because, jaundice comes on frequently towards the close of the fever, without producing any danger; nay, it is rather a favourable crisis; and in cases where recoveries have happened, in the yellow fever, as it is termed, the tinge in the skin has continued for a long time without any other of the symptoms which usually characterize, or more immediately attend, jaundice. Dissections have not shown, in fatal cases, any great derangement in the biliary system; no calculi and little preternatural distension or obstruction; besides, the yellowness comes on very suddenly, and to its fullest extent.—There is indeed a gradation as to the places where it begins, before it spreads universally; the progress of it, however, is different from jaundice. The vessels of the eye, give generally the first alarm of that disease; and the onyx of the nails, becomes very soon affected; but in the yellowness accompanying the Remit-

D   tent

tent of St. Domingo, a different progress is observed; the neck, in the course of the jugulars, the cheeks, in an angle from the nose, forming streaks, give the first intimation; yellow tinges pass along the breast and back irregularly, and the feet are often deeply coloured before the rest of the body is materially affected.—Where the vessels of the eye have been much surcharged, so as to resemble the commencement of real jaundice, with other symptoms proper to that disease, the event has proved, in general, more happy. I have thought too, that the shade of yellow which attends the Remittent of St. Domingo is different from that which accompanies jaundice; perhaps this is a fanciful difference; but what I have stated would seem to indicate some peculiar state of the blood, independent of bilious admixture: future observation may throw more light on the subject.

Upon the whole then, it would appear that miasmata, or particles of elastic vapours, rising from the earth, in the condition called marsh, and acted on by a very vigorous sun, are the remote causes of the Remittent of St. Domingo; and that they enter the human system, either by the vessels of the skin or by the lungs, or mixed with the saliva; that
there

there they produce several morbid phænomena or derangements, probably by forming new combinations with the subtile elements of the fluids and solids, and thus become noxious and destructive, by deranging the peculiar organization which produces the living condition. We must lament our ignorance as to the particular mode of operation, but we may observe certain circumstances which strongly influence the issue of the disease or derangement. To mark these is to make some progress. It is true, that I have heard practitioners chatter about diseases with the same facility they spoke of any common incident; they had a cause at hand to explain every symptom, and a remedy, with a string of certain effects to result from it; to use the expression of an eloquent writer, " they seemed " to have been in nature's cabinet council :" but from such men little can be hoped; we are all, it is true, in the twilight of knowledge; we see to a certain distance, beyond which all objects appear confused, and blended, and I would not readily believe him who would assert he saw distinctly beyond the common limits of vision. In the next section we shall continue to take a further view of the fever and its phænomena.

SECT. III.

Plethoric Habit dangerous in the Remittent—Young Men more exposed to danger than others—Women and Old Men escape the Fever—Practical Inferences from these Facts—Scheme of recruiting Men at an advanced Period of Life.

IT was remarked in the second section, that a habit full, irritable, and plethoric, afforded the Remittent of St. Domingo an opportunity of manifesting its utmost violence: I shall now prosecute this opinion, and deliver the reasons which led me to entertain it. I have already observed, that irregular determinations, founded on a diminished energy in the vessels of some part of the system, were the first effects of the application of the miasmata to the human body; the consequent derangements in the organs, to which these determinations are made, form the danger; the violence of re-action in habits disposed to inflammation is always dangerous, either by aiding the general state of disease, or by producing the direct consequences of inflammation, or rupturing small vessels.

I have

I have seen many corpulent people in the West Indies, who have endured the climate well, but had not all the characteristics of the habit I have been describing, as rendering the Remittent dangerous. I have founded my opinion, of such constitutions being particularly unfavourable, on the following considerations :—First, by observing always, that when the fever invaded such habits, that it was rapid, severe, and violent: secondly, by remarking that young men, from the age of fifteen to twenty-five or thirty, the irritable and plethoric period, were more severely affected than those more advanced in life : and thirdly, by observing that old men, from sixty to eighty years of age, seem as it were to renew life, and enjoy health in warm climates, better than in any other situation ; and when they happen to be attacked with fever, it proceeds with little violence, to a happy termination. Women too are exempted from the violence of fever, except in particular instances, where intoxication has produced the irritable and plethoric state. Females then, and old men, who are in a condition of body directly the reverse of the plethoric and irritable, do not by any means run the same risque when attacked by this fatal remittent. It is fair then

then to conclude, as it is moſt deſtructive in the young and plethoric, that ſomething in that ſtate conduces to this fatality. The eaſe with which determinations are made in theſe habits to particular organs, and the violence of re-action, ſeem to me to be the chief ſources of danger. In old men the fibre is relaxed, and a new condition brought on, reſembling the ſtate of youth; and, in fact, they become fat in warm climates, and more cheerful; a new energy is infuſed in their conſtitutions, and life prolonged beyond its expected period. Theſe changes would ſeem to be produced by the relaxing powers of heat, and ſome ſecret vital energy connected with it. Relaxation and irritability, to a certain degree, are the peculiar characteriſtics of the juvenile fibre. How habit, or nativity in particular climates, operate ſo as to prevent diſeaſes, cannot be explained on any certain principles; we can only acknowledge our ignorance, by ſaying, that nature calculates the ſpecies for the regions they inhabit, or that we gradually acquire conſtitutions ſuited to the climate in which we may be placed.

The doctrine I have been difcuſſing would be of little uſe, unleſs ſome practical inferences could

could be made for the benefit of our troops. If I am right in remarking, that a habit difposed to inflammation, from its peculiar circumftances, is in a dangerous condition, when attacked by the fever of St. Domingo; it will follow, that the troops who are to ferve in that country, ought to be made up of men, at a particular period of life: fuch men too, have the advantage of being well difciplined; their morals are eftablifhed, and their paffions calmed; and they are in every way fitted, for that kind of fervice; whilft the younger troops, are left at home to be formed and regulated. Men from the age of thirty-five, to fifty years, are in a condition to act in St. Domingo; whilft our youth would be fpared and fent to garrifons, where they would be gradually inured to heat, and enabled at a future period to undergo the fatigue of the warmeft region. This plan would be of the utmoft importance; the elder part of each corps might be thus drafted, and proceed with fome confidence to their ftation. Men, who in colder climates begin to lofe their activity and ftrength, would in St. Domingo be in fome degree renovated and rendered again ferviceable. By this means the inactive period of life would be rendered ufeful, and the young and vigorous kept at home,

home, where their ftrength could be moft happily exerted. I truft this fcheme will challenge the attention of our Government; it will prevent many untimely deaths; the fervice in every view will derive benefit; it will have men on whom fome reliance can be placed, when duty requires them to act, and our youth will be faved from the almoft certain deftruction of the climate.

When regiments are thus formed with a view to the Weft India fervice, they would be ftill more improved and protected by refiding in warm, but more healthy regions, before they embark for the Weft. Gibraltar would afford them a mild feafoning; their veffels would become habituated gradually to expanfion, and the fkin would be rendered lax and open for perfpiration. A period might thus be put to the devaftation of that climate, and fcenes, the recollection of which fhocks humanity, would in a great meafure ceafe.

Before I quit this fubject I fhall juft remark, that men at the age of forty years, feem to me very well able to fuftain ten years fervice in St. Domingo; and as all our garrifons are relieved in a lefs period, they would poffefs
fufficient

sufficient vigour of habit to go through every toil. In regiments thus organized it might be proper to reward the services of men, who had acted with reputation as serjeants in a warm climate, and give them promotion. The men of character would thus have something to hope for, and their good conduct would meet encouragement in their profession; above all, the West India service would be carried on with success: at present, neither military talents nor numbers are of use; our hospitals contain our garrisons, and the few who carry on duty are languid and convalescent; they are not fit for enterprize or hazard; and nominal armies will never atchieve conquests.

SECT. IV.

Causes which retard Medical Improvement—Surgery and Medicine compared—Of proximate Causes in Disease—Our Ignorance of these prevents Improvement.

I SHALL now proceed more directly to examine the Fever, and endeavour to ascertain its class; but before its particular history is unfolded, and the practice is described, it will be necessary to take a view of the state of of Medical Knowledge. I have founded my practice on our ignorance of Proximate Causes and the positive nature of fever; and it will be necessary to show the real state of that question, before the principles which guided me are understood.

In all medical researches, we have to encounter those obstacles, which render investigation so difficult, in a science, of which the principles have not been hitherto ascertained. The indolent have abandoned a pursuit so arduous, and adopted the reigning systems of the day. The success of the learned and ingenious has by no means been equal to what might be expected from their talents or industry. Medicine, from the days of HIPPOCRATES, has been a system of varied conjecture, which has changed its aspect in almost every age; each century presented to the student novel doctrines, which in their turn made way for others. These changes were influenced frequently by the progress of natural philosophy, especially by chemistry, but more commonly by the ingenuity or caprice of medical professors. It is fortunate, that amidst these fluctuations of opinion and theory, practice was not much disturbed; it held an even course nearly, and though different

ferent views were entertained as to the mode in which medicines produced their effects, yet the same prescriptions and mode of treatment obtained amongst physicians whose theoretical tenets were very different; on other occasions a peculiar practice was pursued by individuals, very opposite from what might be expected from the theory they maintained. This was particularly the case of SYDENHAM, as Dr. JACKSON clearly points out. There can be no doubt, however, that a more philosophical and rational manner of investigation has been pursued by the moderns. Facts have been more accurately observed, and experiments made with more precision; the laws of the living body have been more attentively examined, and the principles of philosophy are applied with more caution to animated matter; theory too is made to result from a careful observation of facts; it is what it should be, an induction, and not a pre-conceived doctrine.

The circulation of the blood, as ascertained by the immortal HARVEY, throws light on many of the symptoms which occur in diseases, and of which the ancients must have entertained absurd notions. The progress of time and accident have put us in possession of

many

many valuable remedies totally unknown to the earlier phyficians; and the wonderful aid of chemiftry is likely to unravel the myfterious laws of the animal œconomy, and throw light on many obfcure phænomena both of health and difeafe. The chemifts too have put us in poffeffion of many active powers, with which the human fyftem may be affected, fo as to produce changes; and the analyfis of various fubftances, has wonderfully extended our knowledge and increafed our power. The numerous difcoveries in anatomy have improved phifiology and furgery; we have more accurate notions of the feats of difeafes, and we can explain more fully the operation of morbid caufes. The intercourfe between various countries, and the cheapnefs of printing, have increafed in a remarkable manner our ftock of facts. The experience of celebrated men, and their obfervations on the difeafes, climate, and habits of particular countries, thus become a general property in medicine, and may be perufed by every one who has induftry or curiofity. In this manner we are furnifhed with valuable materials, which may be examined and compared, and from which we may draw important and ufeful conclufions. The general progrefs of fcience has no doubt contributed to

the

the improvement of medicine; but, above all, the inftitution of focieties and hofpitals has contributed to correct and enlarge medical knowledge over Europe. In fociety, the powers of the mind are called into action, doctrines are minutely examined, the ambition and induftry of individuals are not reftrained by the terror or authority of teachers; opinions are compared, latent facts are brought forward, and the general refult is publifhed for public infpection. The records of celebrated univerfities will bear ample teftimony to the benefit of medical affociations; hofpitals too are great fchools of practice, where numerous cafes enable the phyfician to generalize, and from a multitude of facts to draw ufeful conclufions; it is in fuch inftitutions that medicine may receive real improvement; the phyfician, free from the fetters which private practice generally impofes, and which the moft celebrated cannot fhake off without prejudice to their reputation, proceeds with more boldnefs to experiment and innovation, which, under judicious management, may be purfued without alarming the feelings of humanity. It is to the prejudices of mankind and the fears of practitioners we attribute

tribute the ineffectual routine of phyſicians, who, fearing the loſs of emolument or employment, dare not venture out of their trammels. To this cauſe alone may be aſcribed, in a great meaſure, the ſlow progreſs of practical knowledge.

Upon the whole, however, Medicine has been ſlowly improving from the days of HIPPOCRATES, though by no means in proportion to other ſciences. It may be uſeful to conſider the peculiar cauſes which retarded its progreſs; for we find, both among the ancients and moderns, that men of great talents and induſtry applied themſelves to medicine; there muſt therefore exiſt ſome inſurmountable obſtacles in the ſcience itſelf. Experiment is the ſource from which, in other ſciences, true or definite knowledge is obtained. In order to attain this knowledge, it is abſolutely neceſſary that the ſubject on which the philoſopher operates, remains in a given or known ſtate, or that its modifications and changes be aſcertained by a certain infallible rule. Secondly, in generaliſing our experience from a few objects of any claſs, ſo as to develope the nature of the whole, it is neceſſary that the few we have examined,

mined, comprehend the laws and nature of all that tribe of objects. If this is not the case, no just inference can be made from the few to the many, nor will experience be useful or extensive; but when the nature of a whole class of objects can be precisely ascertained, by experiments on a few subjects of that class, then, the philosopher can extend his conclusions to the whole class, which he may not have individually examined.— Because, experiments, repeated and confirmed on a few subjects, have ascertained the laws of a whole class, whose essential properties, so far as regarded his conclusions, remained fixed and immutable. It was thus that the immortal Newton proceeded, and from the simple laws of gravitation ascertained the complicated motions of the celestial bodies; in this manner other sciences advance more or less rapidly, but with a degree of certainty approaching demonstration.

In medicine, however, although physicians have appealed to experiment, and made conclusions, yet it will appear that their inductions can never be so precise and decisive as in other sciences. When we examine the influence of experience on practice, we find it
general,

general, though the principles which regulate that influence are loosely and inaccurately ascertained. After all the pains a physician may take, in comparing, examining, and discerning in what constitutions agree or differ, he will be liable still to error from the dubious outlines which discriminate different habits, and the indefinite laws which belong to individuals. This reasoning applies to medicine as a science. Surgery is very different. Whilst it prescribes rules for operations, and discusses the best plan for reducing a luxation, or curing a fracture, it is a respectable and useful art, because such reasoning is founded on the almost unvaried structure of human bodies; but when it deviates into medical reasoning, not founded on these simple principles, it degenerates, and becomes less respectable, because more visionary and uncertain. But to return:

If the nature of the bodies on which the philosopher operates, be either absolutely different each from each, or constantly changing, and if one or more bodies do not contain the collective qualities of the whole, experiments made on a few will by no means be conclusive. In such cases human know-

† ledge

ledge will be always imperfect. But thus it is in medicine. No number of human bodies possess in all respects the same assemblage of properties; these are diversified by endless modifications. The delicate nature of the human system, the difficulty and danger of making experiments, the impossibility of ascertaining their precise effects, the mysterious phænomena of life, the action of animated matter, its relations and dependencies, form such a chaos as confound and obstruct research; experiments, which in other pursuits may be extended and multiplied, are here limited; and inductions, made under certain restrictions and conditions, cannot be wholly trusted. The human frame, though regulated by some general laws, which belong to the species at large, is also subject to the influence of peculiar ones, which affect the individual only, and which are not the same, perhaps, in any two of the species; hence an experiment, made on a few individuals, and applied generally, must necessarily lead into error. If animal bodies were guided and regulated by general laws only, and never affected by the peculiarities which belong to the individual, then similar powers, applied to such body, would always produce similar

E effects,

effects, and a juſt induction could be made from a few to a great number, indeed to any extent; but human bodies are governed each by its own laws, termed by phyſicians its conſtitution. The ſhades, however, which mark and diſcriminate variety, are frequently ſo obſcure as to elude the moſt acute obſerver. From this difficulty much confuſion ariſes in practice. Facts remain as ſuch with reſpect to individuals, but are not ſolid foundations of reaſoning in other caſes, to which they do not fully apply, from ſome ſubtile unknown difference in the conſtitution of each, and yet the circumſtances may have been extremely ſimilar. Some diſeaſes, eſſentially different in their nature and cauſes, exhibit phænomena ſo ſimilar, that the moſt ſagacious obſerver is apt to be miſled, and thus the efforts of the phyſician become pernicious or uſeleſs.

Similar cauſes too produce great variety in the effects, as applied to different bodies, according to the peculiar diſpoſition of each. Hence diſeaſes eſſentially different in themſelves, and produced by different cauſes, are apt to be confounded; and diſeaſes eſſentially the ſame, or produced by the ſame cauſes,

causes, are judged to be different. Of the first class, continued fevers afford numerous proofs, and remittents and intermittents are examples of the second. The treatment accordingly must be often improper, from the difficulty of discriminating. Our best medical records do not afford complete histories of morbid phænomena; because the circumstances in which the difference of diseases often consists are very minute, and do not readily admit of description. Language has not epithets sufficiently accurate or delicate to impress subtile shades, which the eye of the immediate spectator can hardly catch. I have known physicians predict very exactly the issue of a disease from the general aspect of a patient, and many minute appearances, which they could not possibly describe in words, so as to make another fully comprehend the foundation of their opinion. The physician cannot follow the plan of the natural philosopher; the latter can multiply his experiments on matter, to make extensive and general conclusions; but the former is opposed in his career by the moral and civil institutions of society. If he descends to the brute creation, and seizing a chain of analogy, transfers his induction from the one to the other,

other, he will be liable to error: the conftitution and habits of the inferior animals are fo different from ours, that no ftrict conclufion can be made from experiments entirely confined to them; they cannot be interrogated as to the effects of the powers the phyfician employs, and our judgment of their apparent feelings muft be frequently erroneous. We apply powers to affect a body, whofe effential properties are not by any means underftood. The nature of the animal fibre, except a few of its phænomena, is totally unknown to us; that elementary conftitution, which gives it fingular and wonderful properties, has hitherto, and may probably for ever elude refearch; and when we fpeak of applying powers, which are to change its peculiar ftate, we talk a language which philofophy ought to reject.

From our ignorance of the effential nature of animated matter, we neceffarily reafon falfely regarding the direct changes produced in it, either by morbid caufes or medicines. Remote caufes of difeafe often elude the power of the fenfes; but when they are vifible, and fubject to examination, as in the matter of the fmall-pox, we know very little of their mode

mode of acting; we remark, indeed, a number of unaccountable phænomena follow their application to the living fyftem, but that is all.

Such are the boundaries which it has pleafed the Author of Nature to affix to our refearches. The effects of medicines then muft be in fome degree vague and uncertain; but medicines are the power by which the phyfician hopes to produce changes, or alter the morbid condition. If, however, the ftate of the animal fibre is unknown, it will be impoffible to modify with precifion the power which is to change that ftate. It is not furprifing then that medicine fhould fo long be a conjectural fcience. The unmarked variety of conftitutions contributes greatly to embarrafs our purfuits. If we poffeffed a fcale, which, graduated like a thermometer, would exprefs the varieties of conftitutions, then might a regulated and ufeful experience be expected; but, upon the whole, when we confider fully the numberlefs obftacles peculiar to medicine, it is aftonifhing what progrefs we have made. It is unreafonable and ufelefs to expect in medicine the fame fixed and invariable principles which refult from experiment in other fciences. Whoever directs his attention to the healing art, muft content himfelf with probability; if he expects to develope or meet certain and immutable principles to guide his refearches or extend his conclufions, he will be difappointed. Let us take medicine as it is, nor look for what in the nature of things cannot be attained : it has no fixed principles as a fcience, nor any pretenfion to demonftrative evidence. The experience of medicine may ftill be rendered ufeful, and the healing art be placed among the purfuits beneficial to the human race.

It has been queftioned, whether, on the whole, the practice of phyfic has diminifhed the fum of human fufferings, or prolonged life. I, who am willing to give to medicine its full rank, believe it may have contributed to both; but if, on a ftrict examination, it fhould only appear to have merely alleviated pain and diftrefs, even then, it has a ftrong claim to the attention of mankind. It is aftonifhing that an art, which profeffes the diminution of pain and difeafe, fhould have, in all ages, received fo little encouragement from government. In the prefent century men have arifen, gifted with acutenefs and judgment, who have greatly diftinguifhed themfelves;

they

they have opened the road to truth, and prefented to the phyfiologift views the moft interefting and extenfive; they have purfued plans of inveftigation, which promife fuccefs, and may ultimately develope the myftic laws and conftitution of life. In this walk, DARWIN and BEDDOES, hold the firft rank; the talents of the latter have been generoufly exerted, to banifh the terrors of the young and beautiful, in the defeat of a tremendous difeafe. Let us hope that fome fortunate genius may yet arife to difpel the remaining darknefs which furrounds us, whofe bold and decifive talents will bear down all oppofition and difficulty, and in the midft of prejudice rear the durable monument of truth.

Having now pointed out the chief obftacles to medical improvement, I fhall proceed to examine our knowledge of Proximate Caufes.

CHAPTER II.

SECT. I.

The Difficulties of prescribing by Indication—Absurd Reasoning on supposed Proximate Causes—Morbid Action constitutes Disease.

IN observing practice, I have remarked, especially in fevers, the vague and fanciful views of prescription, founded on indications. It appeared to me, that without some knowledge of the Proximate Cause, and its mode of operating, we only lost time in combating effects, the source of which was wholly unknown to us. This rendered practice very inert. The physician became either an idle spectator, or interfered in a manner that promised little success, whilst the disease proceeded in its course with little interruption. The history of fevers, from the days of HIPPOCRATES, exhibits only a humiliating account of idle theories and useless systems.

The ancients blended with their doctrines obscure notions from the reigning philosophy, nor have the moderns been much more happy in their investigations. Fettered by a blind veneration for antiquity, as if age could sanction error, they copied the absurd notions of their predecessors. Few of the moderns have any claim to originality; the features of their systems may be traced in the pages of GALEN, and in the writings of ARITÆUS and AVICENNA. The late Dr. BROWN, though by no means a popular character, exhibited to the public the first philosophical attempt of any consequence in pathology; his fate, and that of his labours, have been justly and pathetically described by Dr. BEDDOES; but although his system is by no means free from errors, it is the most comprehensive and enlightened that has yet appeared.

Fevers, however, are still a barrier in medicine, which neither diligence or talents have been able to surmount. Dissatisfied with the present mode of practice, founded on direct indication, let us examine our knowledge of proximate causes; if it appears that we know very little of these, it will also appear that our indications are often ill founded and nugatory.
The

The theory of medicine, though confiderably improved by a better mode of reafoning adopted by the moderns, is ftill very deficient. If it be deduced from a number of facts accurately obferved, if it confifts of the principles unfolded by experiment, and embodied into a general law, then it will juftly apply to the explanation of phænomena; but it happens too often, that theory is affumed without attention to fact or experiment, and forms the bafis of a fyftem, to which every thing is fitted and cemented, till a flimfy fabric is reared, which the breath of truth blows to the ground.

The theory of the proximate caufe in fevers has varied confiderably in modern times: BOERHAAVE thought it confifted in a ftate of the fluids, which itfelf required proof, and was entirely affumed; HOFFMAN imagined the folids only were concerned; and CULLEN, who copied entirely from him, attributed all the phænomena to fpafm. After all thefe inveftigations, we are yet to learn in what it really confifts. It would be ufelefs at this period to enter on the refutation of thefe doctrines; I believe even CULLEN's fyftem, which was certainly the moft ingenious, has now

now few advocates. It may be remarked, in respect to them all, that effects were seized to explain phænomena, which were in themselves links in the chain of appearances which the remote causes produced. The absurdity of fixing on any intermediate link, to account for all the phænomena, is very evident. We should pay little attention to an artist, who, in explaining the movements of a watch, would pitch on any of the intermediate machinery as the source of the whole. CULLEN's theory had gained considerable credit over Europe, when BROWN's system appeared, which, among other benefits, produced not only a more vigorous spirit of enquiry, but an useful scepticism in system. Before this period, the theory and practice were influenced wholly by the Cullenian school; spasm and its cure were in the mouth of every one, and the pupils of Edinburgh retired from college devoted to this orthodox system.

It is not, however, entirely without use thus to form theories; opinions new and singular awake genius to examine, confirm, or reject them; the faculties of the mind are exercised by research, and its powers increased;

ed; truth may be established, or falsehood detected. When many minds are employed in one research, there is at least a better chance for discovery; the different views in which objects are presented render investigation more easy, and the access to knowledge more simple. To be convinced that we are ignorant is a great step towards improvement, and to discover the insufficiency of a theory stimulates a farther enquiry. In such a collision, a light may at length sparkle to conduct us through the obscure recesses which have hitherto concealed truth; false theories, though dangerous as to their influence on practice, have sometimes been useful, by calling into action the talents of eminent men. To the system of Dr. CULLEN we owe probably the work of BROWN, and certainly the essay of MILLMAN. Let us not entirely banish theory. Even when we cannot clear many doubts, we may thus proceed a certain way, and the journey may be happily finished by a more fortunate traveller. Let prejudice be banished from research, let untenable posts be candidly surrendered, nor let us retain ancient doctrines from an improper veneration for antiquity.

Proximate

Proximate caufes have been fought with great eagerness in all phyfical enquiries; but the magic connection which fubfifts between a preceding and confequent effect, has eluded, and will probably for ever elude, our keeneft purfuit. Philofophy marks a chain, or uniform manner, in which effects appear to be connected, and calls by the name of caufe, an effect which it cannot trace higher, for which it has no antecedent; and which is followed by a train of other effects, which in their turn become caufes, and, perhaps, have no other connection with the higheft links than being merely in fucceffion.

When we obferve a chain of phænomena uniformly and conftantly fucceed one another in a certain invariable way, it is cuftomary to place them in the relation of caufe and effect, though by this mode of reafoning we lofe fight of the higheft link we can trace, and attribute all the appearances to an intermediate one, from which we deduce whatever follows. To illuftrate my meaning, I fhall, for example, take CULLEN's reafoning on the proximate caufe of fever. This celebrated profeffor laid hold of fpafm to account for all the fubfequent fymptoms. Now fpafm

is

is itself an individual effect in the train of phænomena which the morbid cauſe produced; for it has been proved very clearly by minute and accurate obſervers, that other evident derangements are preſent at the ſame time with ſpaſm, and are ſometimes known to precede and ſometimes to accompany it. Nauſea, an inexpreſſible anxiety and uneaſy ſenſation about the ſtomach, languor, and debility, are perceptible before any marks of ſpaſm have appeared. Theſe are modes in which the morbid cauſe operates; they are derangements in the uſual functions, and in a great meaſure conſtitute the diſeaſe; but it would be juſt as fair and as good reaſoning to ſay, that languor and debility, or anxiety, was the proximate cauſe, and produced all the other ſymptoms. The truth ſeems to be, that ſpaſm is a ſymptom of fever, in common with many others, but not by any means the proximate cauſe, as Dr. CULLEN imagined. Fever exiſts and proceeds when no ſpaſm can be traced, nay, when there is poſitive evidence that it is not preſent; for there are clear teſtimonies that ſpaſm, or contraction of the extreme veſſels, has taken place without producing one ſymptom of fever; and there are caſes of fever, where a moiſture has

has continued on the skin throughout the whole course of the disease, and where it has had that soft relaxed feel that indicates a free exit to the perspirable matter. If the production of spasm was at all times followed by fever, which it ought to be if it is the proximate cause, every immersion in cold water would create a fever; but the salutary effects of cold bathing, and the little danger from accidental plunging, are strong arguments against this conclusion. I have enumerated these objections to shew the little use and fallacy of seizing, for a proximate cause, an individual effect of the morbid power, which has no other connection with the subsequent phænomena than an accidental precedence.

Let us now examine what we mean by proximate causes, and what we really know of them. A proximate cause is that condition which exhibits the morbid phænomena, and without which they could not for a moment exist; it is the final operation of remote causes concentrated so as to produce disease and derangement. No part of medicine is so obscure as this; we are daily baffled in our plans of cure founded on indications, because we reason falsely, and proceed to practice

tice on principles not established, and altogether unknown. I am not acquainted with one instance in which we distinctly ascertain the nature of the proximate cause; I allude more especially to fevers; we remark, indeed, its mode of operation, and the phænomena it produces, but the peculiar state necessary to give it vigour, and constitute its essence, is totally unknown to us. When we speak then of proximate causes, we speak of unknown powers producing effects which we observe, and operating in an unknown manner, without being able to ascertain the precise condition which exhibits them. These causes are evidently modified, but we are ignorant of the precise and definite modifications. This being the case, the practice of prescribing by definite indication must be erroneous or feeble; for if we do not know in what the proximate cause itself consists, how are we to prescribe means for its removal? and if it be not removed, we do nothing on this scheme of management. An indication is that method which the operation of the proximate cause points out for its own removal. Indications of cure are always supposed to be founded on a knowledge of the proximate cause; they are the obviating schemes which

we

we adopt to frustrate the movements of a noxious power; but if our chief attention is directed to a partial effect, the disease is permitted to exert its full strength, and to gain such vigour as not to be readily overcome, by any means we can afterwards employ.—Whilst CULLEN's theory, guided practice in fevers, the chief object was to overcome spasm; and medicines were employed for this purpose, which had no other effect than gradually to debilitate, and render the course of the disease more insupportable; by adding nausea to the catalogue of symptoms. It is true, that we are sometimes successful in practising by a supposed indication; when the indication itself is at least doubtful, and certainly the manner in which the medicine fulfils it. Thus, when a sharp instrument or rugged thorn, has penetrated the softer organs, a locked jaw is sometimes the consequence, after the offending body is removed. On the supposition that this arises from extreme irritability, opium is prescribed, which sometimes happily removes the danger, though the precise manner in which the locked jaw is produced, cannot be ascertained, nor the operation by which opium removes it. The precise state then, which necessarily produces

F and

and exhibits the morbid phænomena, being unknown, it is impoffible to form judicious indications, founded in fact, on a fiction of the phyfician's. Till the laws of animal nature are more minutely unfolded, we muft fpeculate; and try to enlarge our views in practice. — We have feen that the practice, efpecially in fevers, though influenced by various theories, has not, for a period of two thoufand years, materially improved. Here then the field is left open to innovation; nor has experience, as Dr. BEDDOES happily expreffes it, any pretenfions to fet up to overawe fpeculation. Where, however, an experience is broad, uniform, and extenfive, it may form a guide, which may be followed with little danger; but where experience leads to no ufeful or decided plan, and where indications are only formed to amufe the practitioner, it is then fair and juft to take fome other ground, till a pofition is difcovered from which we may fuccefsfully play our artillery on the difeafe.

Inftead then of looking to fpecial modes, in which the proximate caufe operates, let us direct our attention to its general effects. If we really knew, in any one inftance, the direct precife circumftances, which conftituted

the

the proximate cause, I am perfuaded the Materia Medica would furnifh us with means to vanquifh it. It is abfurd to fay, that we know the proximate caufe in fevers, whilft thefe fevers baffle our fkill; becaufe, in moft inftances, where we form opinions approaching truth, on the caufes of difeafe, we inftantly fucceed in the cure. The living body, in the ftate called health, performs its functions with eafe and harmony; every part of the fyftem acts in unifon, and agreeable to its nature; producing pleafurable fenfations, and performing every operation neceffary to preferve the whole in perfect order. This harmony of animal action conftitutes good health; it confifts in a peculiar mode of action inherent, or proper, to the feveral organs which compofe the body; but there are powers, or caufes, which feduce thefe organs from their obedience to the proper animal laws, and oblige them to deviate into other modes of action, which create derangements, pain, and uneafinefs, and which ultimately deftroy the fyftem entirely. Thefe aberrations, from the ufual movements of the animal frame are termed difeafes, and the caufes which produce them are morbid powers. The new manner of acting introduced by the caufes of difeafe has

has been termed morbid action.—This term was first used by the late celebrated Mr. JOHN HUNTER, whose original and masculine turn of thinking, introduced many new and useful hints into medical and surgical reasoning.

Morbid action is the derangement in the usual functions, produced by the proximate cause.—As we do not comprehend what constitutes the nature and essence of that power, let us try to modify and change the state of the body; so as to render the operation of the proximate cause less destructive. If we succeed in changing the given state of the body, we assuredly change all the nature of the morbid action, so as perhaps to give rise to a new series of phænomena less dangerous than the former.—It has been remarked, that before morbid powers can produce their effects on the body, there must exist between them what Dr. JACKSON calls an aptitude. This opinion is countenanced, by observing, that though men may be exposed to morbid causes, yet it often requires a long time before disease is produced; that is, before the peculiar aptitude takes place, which disposes the system to yield to morbid influence. In our attempts then, to change the state of the system at once;

it is possible the aptitude itself may be destroyed, and the very principle of the disease banished. The influence of habit is most powerfully felt in all the actions of the living system; hence, perhaps, the state of health is so long continued, and is more natural to the constitution. Whilst the movements of the body are harmoniously performed, and possess vigour, it will be more difficult to impress changes; hence a state of vigour is at all times a kind of protection from contagious diseases; but when the actions of the living system are performed with languor and debility, from whatever cause, morbid changes are more readily impressed and adopted; the feeble rivulet may be diverted into any channel, but the vigorous torrent pursues its course, insensible to small obstacles; hence a state of debility renders the system more obnoxious to contagion, or the influence of disease. I am inclined to believe that morbid causes fail in producing disease; not from the want of aptitude, but of vigour in the contagion, or power itself. Many men, for instance, resist the influence of ardent spirits in certain quantities, while others are easily intoxicated; but every man can be overpowered by a sufficient quantity. Those men, commonly called robust, are not always poſſeſſed of the greateſt animal vigour; ſo that this reaſoning is not contradicted by ſeeing what we call ſtout men readily overcome by contagion. If the phyſician ſucceeds in changing the condition of the body, the whole operation of the proximate cauſe will be alſo changed. This, perhaps, may be called a random practice; but it is not more ſo than that founded on indications; and in varying our means, accident may give riſe to diſcoveries, analogies will be ſeized, and experience conſulted, whilſt the views in practice are enlarged. In an Appendix I ſhall endeavour to ſhow, that this doctrine has ſecretly influenced the practice of phyſicians, without being acknowledged, and that, in fact, the cure of ulcers, as pointed out by Mr. JOHN HUNTER, was directly founded on this doctrine, as well as the mode of treating intermittents and other fevers.

SECT. II.

Opinion of the Yellow Fever—It appears to be the common Remittent—It is not the Fever of Philadelphia *or* Boulam—*The Yellow Fever not contagious—Grounds for this Reasoning.*

I SHALL now proceed more immediately to the object of this work, the Fever of St. Domingo.

After all the instances of this Fever which I have witnessed, and all the attention I could pay to it, I am of opinion, that it is the common remittent of that country, rendered formidable, by being applied to the English constitution; that the variety, which appeared in its progress, depended entirely on the variety in the several constitutions which it attacked; and that the yellowness, which gives it a peculiar name, only marks its worst stages, and is rather accidental than peculiarly characteristic.

Dr. JACKSON, in his treatife, which contains many valuable remarks, has, with uncommon fidelity and accuracy, noted various fpecies of the Jamaica remittent, which feems to me to have been of the fame kind with what raged at St. Domingo, differing only in violence.

Perhaps the immenfe mortality which has happened in the Weft Indies within thefe four years, is to be attributed to the greater numbers who have been fent to that quarter for the purpofes of war; for, befides failors and foldiers, war creates room for a great number of fpeculators; who follow the army from views of commerce. It muft be admitted, perhaps, that the climate itfelf has changed, and has been more injurious to the European conftitution, within this period, than at any former time. What the fecret caufes of this change may be, we do not know; but it has been remarked in the Weft Indies, that during thefe feafons there has prevailed greater heat, and a lefs fall of rain at its proper period; and I have before remarked, that this circumftance always renders the miafmata more vigorous and active; befides, the climate of the moft healthy regions undergoes

frequent

frequent changes, for which we are by no means equal to account; many difeafes make their appearance fuddenly in fuch places, without our being able to explain, in any fatisfactory manner, the means by which they are produced, and they again retire without any evident change in the climate in which they arofe, fo minute are the circumftances which influence the origin of difeafes.

The fever of Philadelphia, which Dr. RUSH has defcribed with his ufual accuracy, certainly never appeared at St. Domingo, during the period of my refidence; though there are many fimilar features in the remittent of St. Domingo, both in the fymptoms and treatment. One important and ftriking difference takes place between them; the fever of Philadelphia was remarkably contagious, whilft that of St. Domingo in no one inftance manifefted that tendency. It is true, that troops have been difembarked at the Mole, and at other places, with a contagious fever amongft them, which carried off numbers; but its type and fymptoms varied confiderably from the remittent. This latter fever appeared in many of the tranfports, who had carried it with them from the encampment formed in Ireland

Ireland previous to their embarkation. From the change and irritability created in all human bodies by the action of heat, the type and form of fevers muſt be changed in warm climates.

There are few caſes of fever, where the pulſe is increaſed, but the hepatic ſyſtem ſuffers ſome degree of repletion, and conſequently its ſecreting powers are increaſed; hence a degree of jaundice is generally complicated with every febrile complaint in the Weſt Indies; but this tranſitory yellowneſs differs very widely, in my opinion, from the inſtantaneous one which takes place in the remittent of St. Domingo. It is true, that numbers ſuffered from a contagion they carried aſhore with them from the tranſports; the 96th regiment were almoſt annihilated by a fever of this deſcription; and other regiments ſuffered alſo from the ſame cauſe; but contagions muſt very ſoon ceaſe and diſappear in a hot climate. The principle of contagion muſt conſiſt in diſtinct elaſtic particles, or be aſſociated with moiſture, or attach itſelf to wood, walls, cloathing, &c. from which the action of heat detaches it ſo as to be applied to the human body in an

active

active state; it must be evident then, that, *prima facie*, warm climates are unfavourable to the spreading of contagions; for the action of heat expands, and rarefies, and volatilises all matter capable of evaporation, and by thus blending them with the atmosphere, either alters their qualities entirely or renders them less noxious. Persons, to receive infection, must in general be very near the source of it, so as to be impressed whilst it possesses vigour, otherwise it fails of effect. We remark further, that all the means we employ to purify chambers, hospitals, or ships, and banish infection, are nothing more than creating an artificial warm climate to rarefy the atmosphere. From this arises the benefit of fumigations, which, perhaps, are only useful in proportion to the volume of smoke which issues from them. No one, I presume, will pretend to point out any new combinations, by which the principle of contagion is neutralized or rendered inert. Of other means of preventing infection I shall speak more fully afterwards.

The Remittent of St. Domingo bears no analogy to the fever described by Dr. CHISHOLME, and which he supposes was carried from

from Boulam, in Africa, by a Guinea ſhip. Dr. Rush very clearly proves, in his own perſon, that the Philadelphia fever was remarkably contagious, and he merits the higheſt praiſe for his fearleſs induſtry amidſt ſo many dangers. Contagious diſeaſes are marked by a ſtriking and rapid progreſs, from a certain point, in which they have commenced, and from which they extend, without any diſtinction, to all around them; when they have found admiſſion into a particular diſtrict, or family, they lay them waſte; and thoſe who are moſt forward to perform the offices of humanity are unhappily the firſt to ſuffer. The friends and attendants of the ſick become infected, and periſh; the phyſician himſelf, more dauntleſs from habit, is at length ſcared from his office, and flies the dreadful ſcene. This miſery Philadelphia, in common with Aleppo, has experienced; but no circumſtance attending an infectious fever occurred in the remittent of St. Domingo. When a ſoldier was ſeized in the barracks, it was not obſerved to ſpread in that particular quarter, and ſometimes only one was attacked; nor could we remark, when they came into the hoſpitals, that in one caſe whatever the contagion was evident. The medical gentlemen could not have

have possibly escaped if there had been any infection; for though they might for some time resist its influence, it is not probable that they would always escape, exposed to the streams of contagion which must have issued from such a number of bodies in its most vigorous state of action; but the gentlemen most exposed to this danger never suffered, so as to suspect that their disease arose from infection. Dr. Scot, Mr. Warren, Mr. Buckle, and many other gentlemen, who gave the sick the most assiduous attention, escaped this fever, although each of them seldom visited less than seventy or eighty patients three times a day. It is true, Dr. St. Clair and Mr. Powrie died; but they had been for a long time exposed to the causes of the remittent before they were attacked. Dr. St. Clair was full and plethoric, and by no means a good subject for any febrile disorder; and Mr. Powrie had been exposed to considerable fatigue. Nor could we remark, in any instance, that the immediate attendants of the sick suffered more than others. The soldiers, who performed the office of nurses, were in general very healthy, and without any fear of contagion. At first, in the ward consigned to my care, I separated the feverish from the others,

others, as much as my limits permitted; but on other occasions I was obliged to blend them with the convalescent; but I never observed, that in this situation any of their immediate neighbours suffered, or that the fever spread. I was led to consider the disease contagious, by reading the publication of Dr. Rush; whose authority must have great weight in all medical opinions; but the fever he so ably describes, differs greatly from that of St. Domingo. This difference may arise from the climate of Philadelphia, and the variety produced in the constitution; but our knowledge is too limited to explain precisely the operation of these causes.

There is no point on which I am more decided, than the absence of contagion in the remittent of St. Domingo. The uncertainty of medical reasoning, and the loose principles on which it is founded, has given rise to a variety of medical opinions on almost every subject; but on this question we were all agreed; no difference of sentiment, no variety of opinion appeared amongst us. Dr. Wright, who was my colleague, and whose accuracy of observation and strength of judgment entitle him to attention, was of the same way of thinking; he had made his conclusions at Cape Nicholas Mole, before I had the satisfaction of meeting him; so that we could not have biassed each other. Dr. GorDon likewise, who had extensive opportunities of observation, and was anxious to ascertain this question, entirely coincided in the same opinion; and if I recollect, it was also the decisive opinion of Dr. Scot; in short, I never conversed with any medical gentleman at St. Domingo, who did not form the same judgment. I had not an opportunity of conversing with the Jamaica practitioners on this subject; but I have been informed, that on several occasions fevers brought there in ships spread for a little time with great severity. This one feature then greatly distinguishes this fever from that which raged at Philadelphia, or the disease described by Dr. Chisholme.

This is a question of the utmost importance to ascertain. If it really was proved, and the proof could not be difficult, that there existed a contagion, our practice and precautions must be different. If there is an infection, it would be useless and inhuman ever to send any European to that climate. Already have groundless fears, terrified and subdued our countrymen;

countrymen; and rendered them more liable to fever, and more easily conquered. The name of St. Domingo is execrated, and dreaded by all descriptions. The officer and soldier bound for this service look upon themselves as doomed to certain destruction. The soldiers lose the benefit of their comrades attention; the officer is approached with fear by his friends or servants; all the soothing attentions, so pleasing in the sick bed, are banished by terror. The service suffers by these false alarms, which are exaggerated in every narrative; and conveyed in the language fear always supplies. The climate is, no doubt, sufficiently terrible to the young and vigorous; many have perished, and will always perish, at that period of life. If the plan, however, of sending men from the age of forty-five, to fifty years, is ever adopted by government; this mortality from climate will in a great measure cease, and the service will be carried on with more vigour and success; at any rate it will be pleasing to know, from the united testimony of all the physicians and surgeons who served at St. Domingo; that the Remittent of that island, called *the Yellow Fever*, IS NOT INFECTIOUS.

SECT. III.

Causes of the Frequency of the Remittent—Similar Causes produce Remittents and Intermittents—State of the Organs in the Remittent—Phænomena in the Remittent—Critical Days—Causes which excite the Fever—Diagnostic difficult—Prognostic.

IT will now be asked, how came this fever to be so frequent, and destructive; if it is not contagious, and what description of fever it really is?

It became destructive, by having a number of strangers presented to its poison; in a condition, unfavourable to their safety. This condition has been already explained, in a former part of this work. The cause which remotely produces this fever is floating in the atmosphere, and breathed by every one alike, or otherwise applied to their bodies. Hence great numbers are seized at the same time, because great numbers are exposed, to the perpetual action of a very powerful agent. This gives the fever an appearance of infection, when it is only the operation of a cause generally

rally acting. Every man is expofed, and probably charged with the miafmata, though we do not know fully the circumftances abfolutely required to make them active. Many ingenious, and ufeful remarks are made by Dr. RUSH on this important fubject. I agree entirely with him, that the caufe of fever may lurk for a long time inoffenfive, till the abftraction of ftimuli, or the addition of them, or accumulated excitability, give them an energy and action. As great numbers are expofed then, many muft be fo impregnated, as really to become feverifh; and as the condition in which they are attacked is unfavourable, the iffue is frequently fatal.

I have already faid, that I think the St. Domingo fever, commonly called *the Yellow Fever*, merely the remittent endemic of the ifland, applied to the Englifh conftitution in a certain condition; and further, that the yellownefs, for reafons already alledged, is not always bilious, but an accidental variety, marking only its worft ftage, and depending on a change in the ferum.—I fhall now more fully give my reafons for this opinion; if they are groundlefs, I fhall think myfelf happy in an opportunity of changing them, for the more

more enlarged and correct views of others, who may be more fortunate in their investigations.

I have observed this Fever, with all the attention in my power; and I have seen it proceed to its fatal termination, in numerous instances, without the least yellowness whatever. Whilst, on other occasions, it made an early appearance, and excited just alarms for the patient; but when it went on without the yellowness, the same symptoms and movements took place, as when the yellowness was present, except the absence of that formidable symptom. The yellowness, if it really marked a peculiar disease, would have along with it peculiar symptoms, which would give character to it, and regularly attend its progress; but no such symptoms ever appear. The incessant vomiting is a symptom common to the remittent in both stages, with, and without yellowness. The mode of attack is precisely the same in both forms, and when we succeed in either, the form of practice is the same. There is no separate and distinct type to characterise a new fever, different from the prevailing endemic. The yellowness, appears to me, to mark only an aggravated cafe of the Remittent; to be merely a ftage of it more replete with danger. The power of the caufes produces more manifeft changes or derangements; and in whatever manner they are effected, they cannot exift without the greateft danger to the fyftem. There does not exift then a peculiar fever, meriting the name of Yellow Fever, in St. Domingo; it is only a variety, marking great danger, and, in fact, nothing more than the common Remittent.

The fever of St. Domingo I have termed a Remittent, becaufe its type refembles that form more than any other. The remiffions, in moft inftances, are very obfcure, and in many not altogether difcernible; I have, however, marked them very diftinctly in a few cafes. They occur, in general, towards noon, and are of more or lefs duration, according to the feverity of the attack; fometimes, however, they are protracted, and happen in the afternoon; but in general fome flight alleviation of the fymptoms, fome relief to the oppreffion, fome diminution in the heat, or in the violence of reaction, are perceptible in the forenoon, and therefore the fever may be called remittent.

I have

I have for a long time thought, that Intermittents and Remittents arose precisely from the same causes, and only differed in form, as the causes were applied to different constitutions, or as different degrees of the same disease. There is hardly a country in which Intermittents prevail, but where also the Remittent makes its appearance. The mild and confluent small-pox every body allows to arise from the same source; yet the appearances are so different in different constitutions, that they would almost seem different diseases. The Intermittent appears to me only the milder form of Remittent, which in itself is the aggravated stage, as the confluent is, of the mild small-pox. When the cause is not very powerful, or applied to a constitution not disposed to adopt morbid movements of any duration, an Intermittent is produced; but when it possesses energy and strength, or is applied to a constitution ready to revolt from the laws of health, and adopt new movements, then the remittent type is completed. In the milder shape of the Remittent, the same remedies effect a cure; and the more the Intermittent approaches, by having no distinct intervals, to the Remittent, the more difficult and dangerous the case becomes.

The Intermittents and Remittents are generally inhabitants of the same country, and either prevail at the same time, as forms of one disease, or appear to succeed one another, from minute changes in the climate or constitution. I have frequently seen the Intermittent commence the attack, and repeat its form for one or more paroxysms, and afterwards, as the cause gained strength, assume the remittent shape, and prove fatal. On other occasions I have observed the Remitting type at once begin the disease, but mitigating in its progress, either from some change in the atmosphere or constitution, assume the Intermitting shape, and the patient has escaped. They have thus appeared in the same places, and have assumed their respective forms, as varieties of one disease, so as to induce me to consider them as only forms of one fever.

The Remittent of Saint Domingo, attacks at all seasons; but with more violence and destruction, during the months in which a vigorous exhalation is going forward; and when the falls of rain are less frequent. From the beginning of May, till the middle of November, the Remittent continues its ravages with unceasing violence; but when the rains fall plentifully,

plentifully, and the heat is somewhat diminished, the Intermittent form begins its reign. They generally commence their attack, either in a state of indirect debility, or where there is considerable excitement. The Remittent usually attacks by lassitude, and weariness, or by chilly fits, and slight pains in the bones, with great inclination to sleep, and an unaccountable listlessness to every thing around. At other times it is ushered in by a regular paroxysm of ague, which, going through its common course, leaves the patient languid and weak; in this state the Remittent assumes its proper form. The pulse, at times, is little altered, and no great change in the heat of the body; but the eye has an expression of anguish, sometimes of ferocity, and a certain grimness takes place in the countenance, as Dr. JACKSON has remarked in the fever of Jamaica. In some instances, the pulse is oppressed and contracted, and the patient is under the influence of very low spirits, and inclined to sigh; in others, the pulse at once is hard and full; the face flushed, and the patient complains of intense head-ach. These several modes of attack are not uncommon. The patient continues in this state during the night, and at times enjoys a calm sleep, at times suddenly starts; and forgetting where he is, sees himself assailed by dreadful phantoms, and wishes to rush into the street, or jump through his windows. When recollection returns, he usually falls listless, or sullen on his bed; and, sighing, sleeps again. During this time all the secretions are considerably disturbed; the urine is in small quantities, high coloured, and turbid; perspiration is irregular, interrupted, and in small proportion; the saliva becomes viscid, and the tongue is covered over with a crust of various colours; the bile is secreted in unusual quantities, and thrown into the stomach, from which it is again speedily ejected; the skin becomes absolutely impervious, and feels like a board; no impression can be made on it by any plan of relaxation, or by any stimulants we yet know. On the second, often on the third day, the dangerous determinations to the vital organs begin; the stomach is assailed, and its coats affected with inflammation; the vessels of them become distended with an unusual quantity of blood, which throws them into an inordinate action, and gives them all the irritability of inflammation; the whole inner surface of the stomach may in this state be considered as one inflamed

inflamed surface, to which nothing is applied with impunity; the vessels, thus distended and active, secrete more copiously, and their secretion is poured into the stomach, which acts with violence to return it; and thus supports a constant determination to itself. At length the vessels, overcome by perpetual action, lose their tone, and pour out portions of blood, which, mixing in the stomach with the former secretion, and an addition of bile, create what is termed the black vomiting, a most dangerous symptom; because the state necessary to produce it, is a state of the greatest derangement. There are proofs of this progress; the pain and irritability of the stomach, and the great secretions in its cavity, argue, in the most decided manner, that the blood vessels are surcharged, and in a state resembling inflammation: that this is really the case, appears from dissections, which show the inner coats of the stomach peeled off, and separated. This could not happen without organic læsion; and such læsions are commonly the result of previous inflammation, and increased action. In this manner is the incessant vomiting accounted for, on pretty certain principles. That this is really the case, may be further argued from

§ the

the state of the skin, it being found completely locked, and shut up, refusing a passage to its most essential and customary discharge. The urine, in common cases of disease, is increased when the perspiration is diminished, and a balance is supported between them; but this does not happen in the Remittent; for though the perspiration is almost entirely suppressed, the urine seldom suffers an increase. The mass of blood, in these circumstances, must be augmented by the retention of different secretions; the consequence must be, that the weaker or more lax vessels will be surcharged, and suffer all the consequences of inflammation. The liver, the stomach, and the brain, possessing a large system of vessels, in a soft medium, become particularly liable to these determinations; and accordingly we find, that in these organs they really take place. In some instances the patient, from the very first moment, feels only a kind of insensibility; and languishes away his life without any pain. The powers of life, attacked in their very principle, yield gradually to the irresistible oppression, of the morbid cause; whilst the system, unable from the beginning to make any proportionate resistance, surrenders itself to dissolution without a struggle.

During

During this progress, changes seem produced in the great mass of the blood itself: what oozes from the gums exhales the most fœtid odour, and the many spots, which, under the title of vibices, or maculæ, are dispersed over the body; argue some considerable change in the solids and fluids. From the fætor of the breath, and the horrid smell of every matter issuing from the sick, I think it will be difficult to question the existence of a putrid state. We see that in the small-pox, a matter often destructive to life is introduced with impunity in numerous instances; and I can see no reason why the putrefactive state may not exist, in a certain degree, whilst the living phænomena are going forward. If it be not, a putrefaction in the fluids; we are yet to learn, what it is that produces that fœtid smell; whilst the blood, by issuing from the gums, nose, and anus, seems really in a more fluid state. A laxity of the solids alone will not explain the hæmorrhage, without a change in the blood itself; and should we admit, that laxity sometimes accounts for the flow of blood, we shall be still in the dark as to the fœtor. It may be proper to remark, that I have frequently seen the dying in a

situation

fituation I could not approach them, from the very putrid smell of their bodies; and that, immediately on their death, they were insufferable, and tainted the air to a considerable distance. The appetite is entirely gone, but when in any degree present, becomes extremely whimsical and capricious. The desire for drink is often remarkable; but small portions only can be swallowed at a time; and these, unfortunately, are again thrown up with violent exertion. The Remittent is at times ushered in with convulsions, which I have seen repeated at the periods of exacerbation. About the third day, sometimes on the evening of the second, or perhaps as late as the fifth, the yellowness begins to make its fatal appearance in streaks along the cheek, forming angles with the alæ of the nostrils; they pursue the course of the jugulars; the back is also tinged in the same irregular manner; the first streaks extend, and become more apparent; the vessels of the eye are evidently affected, and in a few hours the whole body assumes a golden hue; the black vomiting increases, and becomes darker; the patient feels at once relieved, from the pain in his stomach; talks of his happy sensations, which, alas! are only

delusive

delusive preludes of his death. The pulse flutters, and becomes feeble; cold sweats break out on the face; the extremities become cold; the eye, inexpressive, and half closed, sinks in the socket; the pulse entirely ceases, breathing becomes laborious, and the rattle in the throat, announces the near approach of dissolution, which a convulsion generally closes.

I have seen cases where a total insensibility has continued for several days, whilst the pulse supported considerable strength, attended with active hæmorrhage from the nostrils, without affording relief; and yet the patient has recovered. One case of this kind I attended with my friend, Dr. WRIGHT, where these symptoms proceeded for several days in the manner above described; but our patient happily escaped.

It was impossible at times to mark any particular critical days, as deaths and recoveries happened irregularly, without any evident election for particular periods. The fifth day, however, the seventh, and the eleventh, appeared in some degree critical, though not by any means in a certain invariable order. I

have

have seen the fever proceed, without any great violence, to the twentieth day, and yet, after all, prove fatal at a time when hopes were entertained of a full recovery. In such inftances, either the patient was cut off by the gradual and flow diminution of animal powers, or a fudden exacerbation has at once extinguifhed life. In flow cafes, the powers of the human fyftem are infenfibly wafted, and when any exciting caufe is applied, there is no vigour left to combat the difeafe. In a ftate of debility in warm climates, there is nothing left to renovate the diffipated ftrength: the caufes of relaxation are continually applied, whilft the body is weakened in all its functions; hence very few complete recoveries, from a ftate of great debility, ever occur, in the Weft Indies ; but in moft cafes, where recoveries have happened, obftructions are formed in many important organs of the human body; the liver, the mefenteric glands, and the veffels of the fkin, are fo obftructed, that their ufual functions are confiderably interrupted; nor are they reftored to their common offices, before a colder climate has imparted general vigour to the conftitution. Thus, the Remittent of St. Domingo is not only formidable in itfelf, but alfo lays the foundation of many

many other diseases, in the end equally fatal. It may be worthy of remark, that before the Remittent assumed the Intermittent type, a dysentery sometimes intervened, but the Intermittent form generally returned, and after going through some paroxysms, ended in obstinate dysenteries.

With respect to crisis, in this fever, it was seldom very evident; sometimes a profuse perspiration, sometimes the return of sleep, an hæmorrhage at the nose, or sudden diarrhœa, put an end to the disease; on other occasions, it terminated in jaundice, which came on by slow degrees, and seemed to remove all the febrile symptoms. I remarked before, that in some instances the patient was relieved at once, by the appearance of an inflammatory spot on a particular finger or toe, as if the cause of fever had escaped by an explosion. This remark was first made by Dr. JACKSON, and I have seen several instances to confirm it. I could never observe any remarkable lunar influence over the periods of accession.

With respect to Prognosis, it forms, perhaps, the most difficult part of our discussion. The eye, the most interesting organ on these occasions, which seems as it were to predict every event,

event, is a difficult study; the minute changes and variations which it undergoes; which impress the physician, though he cannot describe them, are great difficulties in prognostic; language has not words to describe these minute shades; they can only be felt by the beholder. These difficulties will be readily acknowledged, by those who understand the language of the passions, so easily understood, but so difficult to convey in words. From the eye, conjoined with some other circumstances, I generally drew my prognostic, and I was, unfortunately, seldom deceived in my opinions of danger. It must be acknowledged, that I have met with a few cases, of which I had formed a favourable idea, which afterwards proved fatal; but they were few in number, and occurred in my first acquaintance with the disease.

But the Diagnostic of the Remittent is equally difficult; nor do I now know decidedly any clear and precise mark or symptom by which its commencement could be invariably ascertained. The anxiety of friends, and the decisive steps a physician would take to oppose danger, render the science of prognostic of considerable importance; I shall briefly

briefly state the circumstances on which I usually formed my judgment, as to the issue of a case.

The youth of the patient, and a plethoric state, were invariably circumstances of danger. The state of body, in which the patient was at the moment the disease invaded him, influenced my opinion of his safety. If it came on, after the indirect debility of a debauch in wine, and sitting up late, there was always very considerable danger; nor do I recollect almost an instance of a favourable termination, where the fever thus commenced. I remember being once present at the flank mess, on the Hill, at Port-au-Prince, when considerable quantities of wine were drank, and the party sat up very late; my duty required me to leave them at a seasonable hour; but three of the party, were next morning seized with fever, and two of them perished on the fourth day. I think it necessary to be thus explicit on a subject that so nearly interests us all.

When the fever made its attack, after being exposed to great fatigue, and the action of the sun, it was always attended with danger. If the person attacked was habitually subject to apprehenfions of danger, and low fpirits, the iffue of the cafe was rendered very doubtful. If it made its appearance in habits not circumftanced as I have defcribed, the danger, *cæteris paribus*, was confiderably diminifhed. Combining then thefe confiderations with the actual morbid phænomena in the individual, I formed my opinion, which, in moft inftances was correct.

The morbid phænomena, which indicated great danger, were the following: fuch an oppreffion of all the functions at once, as greatly impeded their action; the pulfe being enfeebled, and the ftrength at once remarkably diminifhed. Suppreffed animal movements, and a general careleffnefs as to the event, indicated no favourable iffue; in fact, where the conftitution made no refiftance, and feemed at once, as it were, vanquifhed and fubdued, there was more danger than even in a violent re-action. Becaufe it argued the complete energy and vigour of the morbid caufe. When the patient changed his natural manner of lying in bed, and affumed any whimfical or unufual pofition, it was no favourable fymptom. Sighing indicated danger; it did not feem to arife from meditation on the difeafe, but involuntarily,

from

from congestion about the vessels of the heart and lungs. The fæces and breath being remarkably fœtid was a fatal symptom frequently; nor were hæmorrhages from the nose, if they were repeated, signs of safety. The tongue afforded also some signs to assist the judgment: if it trembled remarkably on being thrust out, it was unfavourable, or if it was covered over with a leaden coloured crust, whilst the edges wore a brilliant red appearance; a brown or bilious crust is not so formidable, especially if it appears loose, and easily separates when touched. The violence of the general symptoms is commonly attended with danger: vomiting, head-ach, great prostration of strength, when long continued, are symptoms of great derangement, and argue an intense disease. The nervous symptom affords many alarming signs of danger. Tremor of the body when moved, with a tendency to faint on slight exertion, justly alarm the observer; the fierce delirium, which proposes heroic action, and raves of battle, is less to be dreaded than the low, muttering, grim, melancholy, which is lost in meditating wrath, without an attempt to move. But above all, the eye affords the best means of judging, in conjunction with the several symptoms already mentioned: a certain pensive sadness in its glances, an expression of anguish unspeakable, a languor in its movement, an inclination to shut out all objects, are signs of the greatest danger, especially when combined with many of the circumstances above stated. But no description will make the physician fully comprehend what has been said of the eye, unless he has watched it at the patient's bed-side. I have seen a physician so inattentive to the circumstances of prognostic, that he has given out, that a gentleman was recovering, and much better, who was expiring as he was relating the story. This is attended with bad consequences, and brings ridicule and want of confidence on the profession. The most attentive observer will speak with diffidence, but he will often approach truth, and be fortunate in his conclusions. Succesful prognostic begets confidence in the opinion and skill of the physician, and proves to the world that he is not inattentive to the phænomena before him. I have omitted to mention, that the features in general, constituting with the eyes what is termed the expression of the countenance, are of the greatest service in prognosis. A countenance little altered in the general expression does not indicate danger, but where the features

tures lose their peculiar cast and character, and have no expression at all, or appear vacant, considerable apprehensions are to be entertained. When the features express anguish, grimness, or distress, of which the patient himself does not openly complain, though they seem printed on his face, there is considerable danger, especially if sighing is added to the catalogue. There is little to be learned from the pulse; I have seen an intermitting one precede a happy crisis; in general, it is more favourable when strong, than even when full, flow, or equable: when the pulse is not much changed, and when that change is to feebleness, the heart is subdued, and its powers and action diminished.

I have now given the circumstances from which, in general, unfavourable opinions may be formed, though they are not in every case positive or decisive; yet from the combination or presence of a great number of them, a very probable judgment may be given.

Having spoken of the symptoms and appearances, on which unfavourable opinions of the patient's fate are grounded, it will be proper to state the circumstances which afford some hope of recovery. The absence of the symptoms already detailed, affords some prospect of a favourable issue. When the disease attacks a person, not particularly plethoric, or weakened by fatigue, or enervated by debauchery, and where there is a moderate action, and the senses entire, he is in a condition to make a successful resistance. If the remissions are distinct, and the secretions not remarkably changed or impeded; if the fever appears inclined to the Intermiting form; if sleep refreshes; if the mind supports its vigour, whilst there is a sensibility to danger, the circumstances are still more favourable. A deafness occurring in the progress of the disease is not an unfavourable symptom. The gradual return of perspiration over the body, especially towards morning, is also favourable. Eruptions about the mouth and face, with considerable pain and inflammation; a brownish thick crust on the tongue, disappearing from the edges, but leaving them of their natural colour, are no unpromising appearances. Moderate thirst, and moderate heat, without that intense, burning feel, many patients complain of, are promising symptoms.

The natural discharge of the fæces and urine, without extreme offensive smell, and of a natural

tural confiftence and colour, leads to a favourable prognoftic. The coming on of jaundice, towards the clofe of the fever, in a gradual manner, is by no means an unpromifing fymptom. The return of moderate appetite, and a defire for acids, in the courfe of the difeafe, I have often found very pleafing prefages of recovery. The eye, and the countenance, preferving a fteady unclouded afpect, animated by hope, and undepreffed by terror or apprehended danger, afford the moft certain affurance of a happy termination.

I am aware, that all I have ftated forms a very imperfect hiftory of the favourable and unfavourable circumftances, which may influence the judgment of phyficians. Obfervations, and an opportunity of recording them, and multiplying them, can alone increafe our ftock of knowledge in this moft ufeful branch of medicine. It certainly admits of great improvement, and forms, perhaps, the moft ufeful and interefting part of phyfiognomy. By this improvement, we might hope to forefee, at an early period, the force of the difeafe, and be enabled to apply fuitable means, before it attained that ftrength, which we could foretell in its infancy. This certainly would be a

great

great advantage; though, I fear, we shall never attain that perfection in it, which LAVATER fondly thinks attainable. We should indeed be perfect, if we could trace in the features, the small-pox lurking in the habit; or the Remittent, before it produced its peculiar symptoms; but a knowledge of prognostic, far more limited, will be very useful to the physician.

SECT. IV.

Two Cases of the Remittent—French Treatment of them—Dispute in Jamaica *on the Nature of the Fever—Practice of the Gentlemen at St. Domingo—Various Remedies examined—Calomel; Blood letting; &c.—Cases detailed ——The Practice pursued by the Author after long Experience—Comparative Success of this Method—Practical Remarks.*

IT is now time to come to the treatment of this fatal disease, which made so many cruel ravages among our troops, and carried on a destruction almost equal to the plagues of Aleppo.

On

On my first arrival at Port-au-Prince, I had few opportunities of seeing the fever; but very soon, a young gentleman, in whom I was extremely interested, had a serious attack; he complained of a pain in his bones, and a very severe head-ach, with an inclination to vomit; and before I had seen him, though I might probably have recommended it at that period, he had taken an emetic, which operated well, but unfortunately excited an irritability in the stomach, which I could never afterwards subdue. As these symptoms continued on the third day, with a full, hard pulse, and he had just landed from Europe, I directed him to be bled, and accordingly he lost twelve ounces, which afforded him sensible relief. He was about twenty-one years of age, rather plethoric, and somewhat timid, from the histories he had previously heard. Anxious to do every thing in my power for this amiable young man, and not choosing to trust myself in the treatment of a new disease, I begged Monsieur PERE, formerly king's physician under the French government in that island, to pay him a visit; he accordingly came, and ordered him large quantities of lemonade, three injections in the course of the day, a warm bath,

bath, and another blood-letting in the foot, in which he placed confiderable faith in making a revulfion from the head. As I trufted to the long experience of Monfieur PERE, in this difeafe, I did not interrupt any part of his treatment. On the night of the fourth, he ordered him a bolus of camphor with a fmall addition of opium. On the fifth he was vifited again, when I found a confiderable degree of coma prefent, and a low, rapid, muttering voice. The circulation was diminifhing, and vibices made their appearance on the neck and back, intermixed with fpots perfectly black. In this fituation, after the tepid bath, I directed four blifters to be applied to his ancles, and the infide of the thigh, and finapifms to the feet, whilft he fwallowed occafionally a little æther and cinnamon water; but all was in vain, for the cafe terminated fatally on the fame evening. In this cafe there was no yellownefs during the whole courfe of the difeafe, nor any diftinct remiffion, except the abatement which followed the blood-letting.

Another young man was taken flightly ill, on board fhip, a few days before, but did not judge it of any confequence, attributing his

his head-ach and lassitude to a long walk he had taken, exposed to the sun. He was about twenty-one years of age, very robust, vigorous, and plethoric. He had been three days complaining, when I saw him. I ordered him on shore; he seemed better for the agitation of the carriage in bringing him to a lodging. He was bathed in the evening, and I ordered him twenty grains of James's powder, and eight of calomel. This medicine operated well, producing a perspiration and several loose fœtid stools. During the 4th, he was tolerably easy, and drank very freely of lemonade. On the 5th, there was a tendency to coma; and the vomiting became at the same time very distressing and incessant. In this situation, though I had formed no plan of general treatment, I applied a large blister over the region of the stomach, ordered him at the same time an injection, and took eight ounces of blood from his arm. The irritation still continued in the stomach; but as the blister produced its effect, the vomiting gradually abated, and at length entirely ceased; the coma was diminished, and the pulse acquired more vigour and regularity. During this process, there was no remarkable heat, and the skin had not that locked feel, so evident
in

in many cases of this disease. He passed the night of the 5th with more comfort than any previous night since he had the attack. On the morning of the sixth, the vomiting again made a slight appearance, and he complained, that whenever thirst obliged him to drink, it gave him great pain, as he felt his stomach beginning to contract, how soon the liquid touched it. This day I prescribed him small draughts of cinnamon water, extremely weakened by dilution; to which were added a few drops of laudanum. These remained on his stomach, and gave him some relief. I begged of him to avoid motion, and to drink as little as he could, for fear of bringing on again the irritability of the stomach. The tepid bath was repeated; but now the skin became impervious, and felt dry and husky; I remarked too, the vessels of the eye becoming tinged with yellow. On the morning of the seventh, the yellowness had become more evident, and had tinged the skin and nails. He felt very much relieved from pain, his recollection was clear and unclouded, and his mind had all its wonted energy; but his pulse was low, fluttering, and quick; he complained of sharp pains in his bowels, and some difficulty in making water. I ordered his belly to be well
fomented,

Sect. IV.] FEVER OF ST. DOMINGO. 109

fomented, and directed him to get an injection with fifty drops strong infusion of opium. Throughout the day, he was very much relieved, but, towards evening, a violent purging came on, the fury of which nothing could restrain; he was every moment up, discharging small, fœtid, liquid stools; whenever he tasted any thing, it seemed to pass through the intestines with inconceivable rapidity. I tried every means I could possibly devise to stop this purging, by directing the circulation to the surface, by diminishing the irritable state of the intestinal fibre, and by astringents, after the manner of Dr. MOSELY. I could not unlock the skin, which resisted the warm bath, and the action of internal diaphoretics; I could not diminish the irritability of the intestines; nor did astringents prove of the least utility. Whilst the disease was thus holding its victorious career, he became, in a remarkable degree, attached to wine, and intreated me, in a manner too earnest to be refused, to let him have some. I had heard of cures performed sometimes from this delusive call, as if it were the voice of nature, prescribing to herself; and I accordingly indulged him with such portions, as I conceived he might bear without exhausting haufting him; he feized the wine with avidity, pouring on me many benedictions for what he termed the only gratification he could enjoy. The difeafe, however, continued to increafe, his fenfes gradually decayed, he paffed his fæces in bed without any fenfibility, and, altogether, became the moft diftreffing fpectacle I had ever witneffed. On the twenty-firft day from the commencement of the difeafe he expired, whilft, in a fit of delirium, he was attempting to get out at the window.

Thus perifhed two young men of great promife; they were the firft I ever attended in the Remittent of St. Domingo.

The yellownefs in this inftance continued to the laft. What furprifed me moft was, the obftinacy of the diarrhœa, which, in violence, exceeded any thing I had ever witneffed. Since that period I have feen many cafes of the fame kind, which lingered on to a much longer period, but generally proved fatal. In the houfe of Mr. DALTON, a refpectable Englifh merchant at Port-au-Prince, many of his clerks and affiftants perifhed very foon after they landed; indeed, as he affured me, me, hardly any escaped, who were attacked. These gentlemen had all been attended by a French physician. From the specimen I had seen of French practice, I was not much inclined to pursue it further; nor did their success in any part of the town justify any one in following them. I saw before me a very vigorous, powerful, and fatal disease, which performed its operations suddenly, and seemed to require the most powerful means to oppose it. Vigour of disease always requires vigour of treatment. I saw in the French practice no power to change the state of the body; I could observe nothing but a temporising system of nursing, and the disease committed entirely to its own course. I applied myself to such books as were within my reach, but they seemed to converse about other forms of fever, than those before me.

About this time, a most illiberal controversy was carried on by the practitioners of Jamaica, relative to the best mode of practice in the Remittent. The object of this dispute did not seem to be the discovery of truth; it became the means of expressing personal resentments, and rival enmities. Such disputes, conducted on these principles, always
disgrace

disgrace a profession, and bring the combatants into some degree of contempt. Public confidence is lost in men, who are declaiming against each other, and supporting opposite systems of treatment in the same disease. The advocates of each system were keen and active to enlist partisans under their banner, and, in the eagerness of party, truth was warped and perverted. The disputants in the Jamaica controversy, in the warmth of resentment against each other, forgot one general truth, that the least reflection might have taught them; they forgot, that in a disease, which attacked so many various constitutions, in a great variety of circumstances, no one, uniform, invariable mode of treatment could possibly take place with any chance of success. There cannot surely be a plainer maxim, than, that as circumstances and constitutions vary, that the treatment must vary also; but the practitioners of Jamaica universally ranged themselves under two banners; the one maintained the particular efficacy of mercury in all cases; the other, with equal ardour, maintained the superior efficacy of blood-letting, and other antiphlogistic remedies. I had the good fortune, before I quitted the West Indies, to meet

meet several reputable practitioners in Jamaica; and could not help regretting, that their talents had not been better employed, than in a virulent dispute, which could not add to our stock of knowledge.

It was difficult to draw any conclusion, from the facts exhibited in this discussion; each party, as might be naturally expected, produced instances of recovery, under opposite modes of treatment, which each attributed to the benefit of their peculiar management. That men recover under very opposite circumstances will not be questioned, by any one who has seen any practice; because, though the general outlines of a disease may be similar, there exist minute shades, which justify a different treatment. The constitution of two patients, under the same form of fever, may be widely different, and consequently a variety will be produced in the effects. Besides, opposite modes of treatment, supposing the cases, to be the same, may cure, because each mode operates a total change in the given condition of the body, and thus banishes the morbid phænomena; so that neither of the systems pursued in Jamaica, derive any great support, from this casual success. If I recollect right, the successes of each seemed nearly balanced,

balanced, if we may rely on the news-paper accounts publiſhed at Kingſton.

Puzzled, and diſſatisfied with theſe accounts, I betook myſelf to the ſtudy of the diſeaſe itſelf; till the phænomena ſhould teach me ſomething of their nature, ſo as to form a mode of treatment. Soon afterwards, I had occaſion to viſit ſome ſailors on board Mr. DALTON's ſhips in the road; they had been ill for ſome days, before I ſaw them, as it is the manner of ſeamen, to conceal their diſeaſes, till they can no longer be kept ſecret. I found ſeveral affected with a ſmart fever, the pulſe quick and tenſe, the countenance fluſhed, attended with a conſiderable degree of head-ach: one or two indeed were in a different ſituation; their countenance expreſſed anguiſh, they ſighed, inſtead of reſpiring, and the pulſe was low and feeble. In one of them, the hiccup had juſt begun. The firſt I directed to be bled pretty freely; and to take fifteen grains of James's powder at bed time, to which an emollient injection was added. They were conſiderably relieved; ſome degree of perſpiration had been produced; and the head-ach was diminiſhed; but they were extremely weak, and in one of them ſome yellowneſs appeared. To the other two I preſcribed a large bliſter each, over the region of the ftomach, with a camphire bolus, and a fmall portion of opium. When I returned next day, I found one of the latter had expired towards morning; and the others were confiderably relieved. They, however, recovered after a long convalefcence.

It was a practice followed at Port-au-Prince by the medical gentlemen, when I firft arrived there, and I purfued it alfo, the moment any one was feized, to order him a tepid bath, to cleanfe, purify, and relax the fkin, fo that there might be no obftacle to the free exit of perfpiration. After the patient was put to bed, the belly was opened by a lenient injection; and eight or ten grains of calomel with a portion of James's powder, in the form of pills, were generally prefcribed, to be taken immediately. Thefe ufually procured the difcharge of large quantities of bile, either by ftool or vomiting. If however this quantity had no effect, which fometimes happened, the injection was repeated, and a larger dofe of calomel, joined to fome purgative, was again given. If they had ftill no effect, the dofes were ftill increafed, till fometimes an amazing quantity of calomel was fwallowed without the fmalleft apparent effect. At times

times a fudden falivation made its appearance, which, in general, put an end to the fever; but which itfelf became a moft formidable difeafe, which nothing could reftrain. It is true, that many have recovered after a falivation was excited, but they are ufually thrown into a moft dangerous ftate of debility; from which they feldom attain any ftrength. One cafe occurred in my own ward in the hofpital, where the patient got entirely well, of the fever, but the falivation refifted every poffible means I ufed, to reftrain it. Mild purges, local applications near the mouth, to divert the circulation, ftrong aftringents, all were employed in vain; it proceeded without abatement till the exhaufted patient funk under it. Dr. Scott vifited this perfon with me, but all our treatment was in vain.

If after thefe prefcriptions, the fever did not abate, tepid baths were repeated, and diaphoretics adminiftered; with mild diluent drinks, and fuch form of nourifhment, as was eafily digefted without giving uneafinefs in the ftomach. Such was rice water, boiled to a certain confiftence, in which the foluble farina, blended with the water into a cream. This cream was mixed with a fmall portion of Madeira

Madeira and sugar, and the patient directed to take small portions of it, as the stomach would bear. If any distinct remission came on, which was rarely the case, the cold infusion of bark, with cinnamon water, or other cordial addition, was prescribed; at other times the decoction. Though I can say little for the efficacy of the bark, except in the Intermitting form. I certainly observed no instance in the Remittent in which I could remark, that it produced any considerable effect at all. Other symptoms were combated as they arose. When the vomiting was severe, blisters were applied over the region of the stomach; injections were prescribed to remove any source of irritation in the bowels, and such drinks were ordered, in small quantities, as were of a nature to produce the least possible irritation. Of this sort, were beef tea, toast and water, rice water, with a little cinnamon boiled in it; and when the strength and pulse seemed to decay, and resist all these means; wine was ordered, and the warmer stimulants, to support if possible the living phænomena. Camphorated bolusses, and blisters to the extremities, were also added to the plan of treatment. Such were the outlines of our general practice,

practice, on my first acquaintance with the disease.

It is to be remarked, that though bloodletting was occasionally performed, it did not become an indispensible part of our plan; for many instances of fever occurred, where bloodletting did not seem necessary; the person being previously weakened, by a long residence in the climate, and not in a situation to undergo any considerable loss of blood. Under this mode of treatment, most of our patients perished; very few, in my opinion, owed their lives to our practice; and yet we bestowed every possible attention. It is not easy to conceive the situation of a medical officer, placed in such a dilemma, surrounded by hundreds of his countrymen, in every stage of misery, looking up to him for assistance, when he cannot rescue them from impending danger, whilst every scheme of practice, every exertion of thought or industry, every experiment fail of success. It is then humanity, to consider every circumstance, that can possibly afford a chance; it is then, fair to deviate into new paths of treatment, which cannot be less successful, than what we have already tried. I have witnessed scenes of misery, I can never forget;

forget; they impressed me too powerfully to be erased from my memory.

It is proper in a question so important, as the treatment of this formidable disease, to be very explicit on every method that has been tried; and to reason on them calmly and fully, before any plan is relinquished that might afford the least chance of success; and also to weigh carefully the grounds on which any new opinions are admitted as foundations for practice: I shall therefore take a view of the success attending the various plans of practice in our hospitals, and, so far as I could collect information, in Jamaica. I must confess, that the plan I shall recommend had not so great a trial, as I might wish; in order to establish it. The opinions and facts on which I ground it, did not occur to me in their full light, until a short period before I left Saint Domingo; so that I had not many opportunities of applying the principles to actual practice; which alone is the true test.

Let us first attend to the plan of giving calomel. The uncertainty of its operation forms in my mind a very strong objection against it; many patients have swallowed some hundred grains

grains without producing any effect at all. Now the rapid progress of the disease, requires something on which we can positively rely, which will produce its effects quickly, and which in all cases will cause some evident change. If, therefore, a medicine is uncertain in its operation, and does not act speedily, it certainly does not promise to be an useful remedy in a disease so rapid, and of so much vigour. The intestines, by some means or other, are rendered very insensible in this fever, nor are they readily affected by any medicine, whether it is that they are covered over with a large secretion of mucus, which guards their inner surface, or that their general sensibility is impaired, cannot be easily ascertained, but it is a certain fact, that whilst the stomach is agonised and convulsed by extreme sensibility, the intestines maintain the most obdurate inactivity. Aware of this, Dr. THEODORE GORDON, if I mistake not, used unction, and tried in this manner to convey mercury into the system; but the absorbents seemed equally insensible, as in the intestines; and when calomel really affected the system, we could not by any means regulate the effect produced, or know what quantity to prescribe, or when it would act. Whilst we were thus amused,

looking

looking out for the effects calomel was to produce, we were often difappointed, whilft the difeafe continued its courfe without interruption; and when it did produce falivation, the index of its having entered the fyftem, that very falivation became a very ferious difeafe, and left the patient in a ftate of dangerous debility. But it feldom happened that it produced this effect with any certainty; it much oftener remained dormant in the body, without producing any evident change. A medicine then, of this uncertain defcription, of which three grains will fometimes bring on changes; when at others, five hundred are taken without effect, cannot be relied on, in a difeafe, where every means we ufe, ought to create fome alteration in the exifting circumftances. But calomel has never fhewn any fpecific power in this difeafe; its great ufe has been as a purgative, and perhaps an alterative of the given ftate of the fyftem. In this view it was ufed, I think, by RUSH, in the fever of Philadelphia. Nor is it without great ufe. It clears the inteftines from the fæces which would otherwife remain, and prove very injurious; it affifts in killing and expelling worms, which are often troublefome; it unloads the veffels, which deter-

minations

minations may have furcharged; and it may contribute to effect an imperceptible change in the exifting circumftances of fever, fo as to render it more eafy of cure; but, fo far as I have feen, it cannot be relied on as a principal agent in the cure of the Remittent.

It is true, that in our hofpital, and in Jamaica, fome recoveries happened where large quantities of calomel had been taken; but it had produced no vifible change in the circumftances of the difeafe. The Remittent went through its ufual revolutions, without interruption, and ceafed without an evident crifis. The movements of health ever ready to return, and more natural, from habit, are at hand, to commence their action; when the morbid action is not vigoroufly fupported by the morbid caufe. This is a fair, candid account of calomel, fo far as I have obferved its effects, or gathered them from the remarks of others.

Let us next examine Bark, as a fpecific, which has been prefcribed frequently in the Remittent, and from which practitioners expected confiderable fuccefs. Whenever the inteftines had been previoufly evacuated; and

the

the Remittent had shewn the least abatement in the severity of the symptoms, so as to merit the name of remission, I constantly prescribed the bark, at first in substance, in doses of a scruple, repeated as often as the stomach would bear it. But even these small doses, were often rejected, and left such a degree of sickness behind them, that I was under the necessity of abandoning the bark in substance. The cold infusion, and the decoction, were used instead of the powder, in as large quantities as the stomach would bear, but even this was rejected. I must say that I never saw any instance, in which the bark decidedly put a period to the return of the fever; except in cases where there was an evident tendency to very distinct remission, previous to its use. In such instances the Remittent is inclined to adopt the Intermittent type, and consequently to assume a form less powerful, and more readily subdued by the bark, which in that shape of the disease, enjoys a specific superiority, over any other medicine. It happens frequently, when the Remittent has ceased for some days, that it will again return, with increased violence, and often on such occasions proves fatal. Nothing is so powerful in preventing such returns as the use of the bark; which ought to be begun, the

moment

moment the remission is distinct. In instances, where the stomach could not retain the bark; the infusion or decoction, I have directed it, to be given by injection; sometimes in powder, to the amount of half an ounce thrice a day; mixed in three ounces of water, and with fifty drops of laudanum. At other times I prescribed the decoction, or the infusion, repeating the injection oftener. This mode of prescription, I found very useful and successful. The stomach was not irritated, nor did the patient complain of that nausea and anxiety so distressing to his feelings. Every effect, that the bark commonly produces, was thus obtained, in an easy, agreeable form, without giving disgust, which too often arises from the repetition of this medicine. All then, that I could see, in the use of the bark, was, that where the fever, showed a disposition towards remission, or actually remitted, it prevented, on many occasions, the return or renewal of paroxisms. I found it most successful in the form of injection, because in this manner I could carry the use of it much further than in any other way. I had no good opportunity of ascertaining, whether the red or pale bark was most powerful; in the few instances, where I attempted this experiment, I could not observe
any

any superiority in the red bark; it did not appear to operate in lesser quantities; nor were its effects more speedily produced. But it would require a more enlarged comparison, between them, to ascertain the fact, and a more sedulous attention to circumstances, than I had the power of paying.

Let us next examine the effects of Opium in the Remittent. This most useful medicine, the kind soother of agony and pain, requires to be used in this fever with the utmost caution. I have found the remarks of Rush on this subject, in the treatment of the Philadelphia fever, nearly coincide with the observations I made in my notes on the Remittent of St. Domingo. In the beginning, I always found it injurious, where restless nights, and anxiety tempted me to prescribe it in large doses. It procured no settled rest; for a time the delirium was increased, to which a stupor rather than sleep succeeded; and the next day languor, irritability, and weakness prevailed; in short, laudanum did not seem by any means to amend the state of the patient. It is, however, a very difficult point to settle, how much of a new stimulus must be given, to do away the action.

action of one already operating in the body. Perhaps we fail in our hopes of opium from this very difficulty; and frequently from giving it in small doses; but the tendency to coma, and the irritability produced by it, hindered me, from carrying the use of it further, than what the agonies of the patient, extorted from humanity. Opium, unless it is pushed so great a length, as to banish the action of other stimuli, can be of no service; it is temporizing with the fever; and lulling the disease asleep, perhaps to recruit its vigour. I have sometimes found it useful, given by injection, in moderating the violence of vomiting; on such occasions, I have prescribed two hundred drops, in a small quantity of warm water. I have joined it also to antimonials and camphire, not so much with a view to produce a general effect in the system, as to secure the retention of the other medicines in the stomach. In this view I found it sometimes useful. I have seen the Remittent in many instances ushered in by convulsions, which were repeated at the periods of exacerbation; in these cases large doses, by the mouth and by injection, produced happy effects, and seemed to leave the fever in a more manageable form. In

cases

cafes too, where remiffions had commenced, and where a return was apprehended, I prefcribed opium very freely; and, as I conceive, with good effects. Towards the happy termination of the Remittent, when the patient was fatigued by reftlefs nights, I found opium of very confiderable ufe; but it was freely prefcribed; and other ftimulants were given, when it was withdrawn from the patient. Upon the whole, opium cannot in our prefent ftate of knowledge be ufed with advantage, in the beginning of the Remittent; but it is of fignal fervice in many occurrences during the difeafe.

Antimonials, under which we range JAMES's powders, did not appear to me of any great fervice; fometimes they produced perfpiration, which afforded temporary relief from the fymptoms; but they again returned, and continued their courfe. In my early practice, indeed till a fhort period before my departure, I was in the habit of continuing pretty large dofes of James's powders joined with calomel; and though in general they produced fome flight abatement in the fymptoms, yet they could by no means be relied on as decifive means of cutting fhort, the courfe of the fever.

In

In one instance, where I had an opportunity of seeing the first evident effects of the Remittent, in a gentleman, who had been a few days landed; I prescribed him twenty-five grains of James's powder, and ten of calomel, after a warm bath: an astonishing sweat was raised, and he had several copious, and bilious stools; there was a complete remission next morning; when he was ordered a scruple of bark every hour, which he continued for some days without any symptoms of a return, and he recovered without any further repetition of the disease. It is probable, that in many instances, we might be thus successful, if we had an opportunity of being called in so early; but it more frequently happens, that the disease has run a course of three days before its aggravation obliges the patient to call for assistance. The first approaches, in fact, are insidious, nor is any one aware of danger; a pleasing languor induces sleep, and a desire to recline, nor is it, till head-ach and pains in the bones arouse attention, that a person thinks himself unwell. At least these were the sensations I felt in an attack of this fever, which soon disappeared. It is particularly difficult, to see the early stage of the Remittent among soldiers, foldiers. They never complain in the beginning, and have fuch an averfion to go into an hofpital, that they conceal their fituation till the difeafe has confirmed itfelf in their conftitutions. It is not furprifing that foldiers diflike general hofpitals; they fee very few return, who once enter their gates.

Having confidered the chief means ufed to combat the Remittent; I fhall now examine the effects of Blood-letting fo far as I had an opportunity of obferving them. Having been for fome years in the habit of feeing the inhabitants of warm climates, I was impreffed with one general opinion relative to blood-letting, which was, that the inhabitants of warm climates, after a refidence of any length, could not afford to fpare blood, from the purpofes of the animal œconomy. To this opinion I was led, by obferving in general the lax ftate of the fibres, of thofe who refided in warm climates; the diminution, or rather the abfence of the rofeate European bloom; and the great wafte of the fluids by perfpiration. Befides thefe confiderations; I remarked, that though the inhabitants of warm climates poffefs great activity; and are more fprightly and lively,

K than

than the Europeans; they, by no means possess the same strength. Whether a certain state of the blood, that we call dense, red, and healthy, be the cause or effect of strength in the fibres, may be difficult precisely to ascertain; but this, we ascertain distinctly, that it is intimately connected with such a state of the body. The state of strength, and rosy colour, are always connected with a dense state of the blood. But the pale relaxed habit never produces, dense and red blood. Physicians have drawn some conclusions from these phænomena. They infer that the first class bear blood-letting, without any material injury, and often with advantage; whilst the second, cannot bear evacuation, without injury to the constitution. Impressed with this doctrine, I held blood-letting in a warm climate to be in general improper; without reflecting, that although this reasoning might apply to those, who had for any time resided there, it could not apply to new comers, who had not been exhausted by perspiration or relaxed by heat. Besides, even in the feeble class, who may have resided for some time in a hot climate; if they are seized with acute diseases, I can see no impropriety in blood-letting; because this evacuation, by removing

ing a state which would destroy the system, must be less injurious, than a temporary debility. I should not therefore hesitate in some states of inflammation, to bleed freely, even in warm climates. Though I would not push the evacuation to the same extent as in Europeans, newly landed.* I am doubtful, whether in all cases of hepatitis, blood-letting be proper; as the inflammation, may be of the passive kind; and mercury seems to succeed in the cure, by its stimulant power chiefly.

From this view of the constitution in warm climates, I was improperly led to consider blood-letting as always injurious, and consequently abstained from it. I found besides, that the gentlemen, who had a longer experience than myself at Port-au-Prince, had not adopted it as a means of cure. The French indeed had bled very freely, at every stage of the disease, but they carried it beyond the proper bounds, and I saw an instance or two, where their patients sunk under this evacuation. The fate of Lieutenant S———, of the 18th light dragoons, was certainly hastened by this treatment. If blood-letting produces good effects, which I believe it does in most instances, it must be performed very early in the difeafe, and be performed with boldnefs. I fhall have occafion to explain this more fully hereafter. When I began maturely to confider the difeafe, and the fubjects which it attacked; when I obferved its phænomena and fymptoms with accurate attention, I then judged that in moſt inſtances an early blood-letting might be uſeful. After I had formed this opinion, the firſt opportunity which occurred of trying it was on my worthy friend Captain S——, of the Royal Artillery, an uſeful and active officer. I happened on my return from another ſick officer to call on him by chance, and found him with a very ſmart fever; his pulſe quick and ſtrong with intenſe heat, an inclination to vomit, and his face very highly fluſhed. Captain S—— was of a florid complexion, plethoric, and vigorous. He breathed with difficulty, and inſpired with a ſigh. There was ſome ſlight delirium, and want of recollection. Having forgot my lancets, I ſent a ſervant for them, and waited by his bedſide, till they arrived. Leaſt he ſhould be alarmed for himſelf at ſuch a meaſure; for it was a late hour, I propoſed it to him, without urging it, as abſolutely neceſſary; he agreed without hefitation; and I accordingly took about ſixteen ounces from

his

his arm. He was fensible of inftant relief; the head-ach abated; the flufh in his countenance fubfided, his breathing became eafy, and his recollection complete. He turned round in his bed, and fell into an eafy, profound fleep. I vifited him next morning; when I found him much more eafy than the preceding day; though the fever had not wholly left him. He had enjoyed a comfortable night, and felt no diftreffing fymptom, except a fenfe of laffitude and wearinefs, infeparable from fever. His tongue however had a bilious afpect, and was covered over with a yellow cruft. I prefcribed him fix grains of calomel, with ten grains of James's powder; which operated in the courfe of the day; and procured the difcharge of much bilious matter. The fever became more moderate, though there was for feveral days a want of recollection, and fome degree of delirium prefent. When the remiffions permitted me to ufe the bark, it was given, and the cold bath completed the cure. Captain S—— was thirty years of age.—The great difficulty in the Weft Indies, was to procure ftrength after the fever ceafed; this was often unattainable without change of climate.

The next cafe in which I tried the effects of blood-letting was in my friend Capt. C———, of the 41ft, whofe exertions at Bizoton, and wherever the fervice required him, at length brought on an attack of the Remittent. It has been rightly remarked by Rush, that the caufe of fever often lurks in the body without being called into action for many days; and I have no doubt, but it may again pafs out of the body, without producing any morbid change; unlefs it be affifted by the addition or abftraction of other ftimuli. Of the truth of Rush's remarks, I have feen numerous proofs. Both Captain C——— and Captain S——— were attacked, after having undergone confiderable fatigue on fervice. They were much expofed to laborious exertion. When I vifited C———, I found him affected with a confiderable degree of fever; he complained of great pain in his back and loins; and an inability to maintain an erect pofition. There was confiderable heat, and fome degree of head-ach. Having a lancet in my pocket, I inftantly bled him pretty freely, to the amount of twelve ounces. I ordered his feet to be bathed in warm water, and prefcribed one fcruple of James's powder, with ten grains of calomel. Thefe produced nearly the fame

effects

effects as in Captain S———'s cafe, and he found himfelf fo well in the morning; that he was imprudent enough to venture out, and ride. The fubfequent ufe of the bark prevented any return. In feveral cafes among the foldiers, I performed the fame operation, and prefcribed the fame medicines, when I had an opportunity of feeing them in the early ftage. If the blood-letting is not performed on the fecond, or at furtheft on the third day, I do not imagine it will have fuccefs. But in the cafes where I had an opportunity of acting on the firft or fecond day, the event was in general fortunate; though fome perifhed; nor is it likely that any method will ever be difcovered, which fhall invariably promife fuccefs. But we do a great deal, if we leffen the mortality.

In another inftance, in Captain L——, of the 82d regiment, a very ftout, plethoric man, whom I faw the morning after the fever had made, its manifeft appearance, I directed a very large blood-letting to the amount of fixteen ounces. The blood, as Mr. BELL, furgeon of the artillery, informed me, was fizy. Capt. L—— felt very confiderable relief from the evacuation; the head-ach abated and his recollection

collection became more clear; he was even cheerful; and though naturally a timid man under illness, spoke with confidence of his recovery. After the blood-letting he had several injections, and took one scruple of James's powder, without calomel, as the injections had operated very freely. The powder had produced some perspiration, and he was in the evening much better. Next day, however, the head-ach again returned, with some degree of delirium; and the pulse became rapid, and strong. I directed Mr. BELL, to take away eight ounces more blood; and repeat the powder and injections. After this, Dr. SCOT and myself visited him late at night, and found him so free from fever, that we concluded, he had every chance to do well. I never saw him afterwards, having been taken very ill myself that night, and rendered incapable of returning to him any more. I understood however, that next day, he had an exacerbation, and vehemently demanded some porter or wine, which being delayed or refused, he was agitated by a most furious passion, after which he sunk into an irrecoverable debility, and expired with spasms and convulsions. This is not the only instance in which violent excitement has proved suddenly fatal.

fatal. I confess, that his death appears to me, to have been brought about by anger only.

To these cures, too few to form any decisive opinion, I shall add, that Dr. JACKSON, whose sagacity and attention are equally conspicuous, was much more successful, after he made more free with the lancet. The 56th regiment had been suffering from the Remittent very severely at the Mole, both before and after they were landed; they had lost a number of men, but when Dr. JACKSON took the management of the sick, and bled more freely; the mortality diminished considerably. And though the fever described by RUSH, differs materially in being highly contagious, yet there are circumstances of great similitude in the symptoms; and I look on Dr. RUSH's practice as a confirmation of the benefits of blood-letting.

Let us next attend to the effects of Purging in the Remittent. As the intestines, from the beginning, are affected with inactivity, it is necessary that they should be roused by stimuli, to discharge their contents. It has been observed besides, that bile is very amply secreted, and at times passes to the intestines, where

it

it would become putrid and ſtimulant; and would of itſelf be a ſource of fever. It is proper then to employ purging to clear the inteſtinal canal from fæces, which, if retained, would prove an unpleaſant ſtimulus, to the whole ſyſtem. In ſuch caſes of inteſtinal inactivity and retention of fæces, it is probable that ſome particles of a putrid nature paſs through the lacteals into the blood, where they cannot produce any ſalutary effect; but contribute to the general irritation then preſent. But purging is a means of diminiſhing re-action, and leſſening the velocity of the pulſe. The increaſed ſecretion from the veſſels of the inteſtines, diminiſhes the abſolute quantity of the circulating maſs, relieves the veſſels from tenſion, and renders the danger of determinations infinitely leſs. Purging too, I conceive in many inſtances may remove miaſmata from the inteſtines, which might add to the ſeverity of the diſeaſe. It is a mild ſpecies of evacuation, which patients inclined to paſs into a low ſtate, bear better than any other. I queſtion, however, whether it be very proper except in the early ſtage; as it may be the means of increaſing the irritable ſtate of the ſtomach. And beſides, as we wiſh if poſſible to avoid determinations,

tions, to the inward organs, it is doubtful, whether purging may not increafe this tendency. In this view, it is better perhaps to employ it early; and in the further progrefs of the difeafe, to truft to injections. Thefe clear, not only the rectum, the natural depot of the fæces; but a ftimulus is commonly applied to the upper inteftines, which folicits them to propel their contents. I have, in fome cafes, placed much reliance on this evacuation; where the patient has been full, but of that habit, which inclines to melancholy, or nervous, and which would not bear the lancet. But I cannot fay I have reafon to boaft of its fuccefs.

On the evening of the 17th June, 1795, I was called to vifit my friend Capt. I⸺⸺, of the 69th regiment. I found him in bed, complaining of a dull heavy pain in his head; with a very flight naufea, and a fenfe of obftruction in the noftrils, as if he had caught cold. He faid, the firft fymptom he had remarked was drowfinefs, and an inclination to recline. The pulfe was frequent, but not hard, and the tongue flightly furred over, with a leaden coloured cruft. The fkin was open, with very little increafe of heat. Capt. I⸺⸺

was

was about thirty years of age; full but not florid; and of a bilious aspect. I directed his feet to be bathed; and gave him four calomel pills, containing in all ten grains. On the 18th, I called in the morning; the general symptoms, much as on the 17th; the head-ach increased; he had vomited the pills, soon after he took them, passed the night rather restless, and with the pills, had brought up a large quantity of deep coloured bile, but had no stools. I ordered him the calomel again, but in smaller quantities, to be repeated, till some effect was produced. During this day he had twelve evacuations; which, as he said, scalded him in the passage; but the head-ach and drowsiness abated; though there was a kind of stupor still remaining. I wished now to apply a blister to the neck, but he positively objected; and said he only wanted something to procure rest; after the fatigue of the preceding night. In the evening the tepid bath was ordered, and he took at bedtime

℞. Aq: ammon. acet. ʒi.
Cinnamon. ʒss.
Tinct. op: gtts. xxx.
Sumend. h. s.

On the 19th, found him much refreshed, after a sound sleep, and a glowing general perspiration; the tongue continued loaded; and there appeared still, a great determination towards the head. There was a kind of insensibility to common occurrence, and a carelessness as to the event, which I have often seen, a presage of danger. I ordered him again the calomel pills. 20th, he passed the night without any sleep, but felt no pain; the pills operated towards morning, and produced seven complete evacuations. The heat was nearly natural, the tongue covered with the leaden coloured crust, but clearing from the edges and tip inwards, leaving them of a bright red colour. The head-ach was gone, the pulse 88 in a minute, rather contracted, very little thirst, the countenance dusky and grim, with a strong propensity to sleep.

℞. Infus. cort. Peruv. ℥viii.
 Tinct. colomb. ℥ii.
 Aq. ammon. acet. ℥iss. M.

Of this he took a spoonful every half hour. During the day, the bark sat easy on his stomach, and he passed a good deal of bilious matter in his stools; but the pulse became creeping and small, and the skin continued relaxed,

relaxed, without moisture. In the evening I directed him, to be washed with cold water; and to have some buckets of it dashed over him; after which he took at bedtime,

    ℞. Aq. ammon. acet. ʒss.
    Pulv. Jacob. gr. xii.
    Tinct. op. gtts. xv.   M.

He was ordered wine and water for drink, in the proportion of one third of wine. 21st. I visited him very early in the morning; and found him very comatose, and the pulse just perceptible. I immediately rode off to the hospital, and returned with three blisters and a camphorated mixture. But I found the agitation of death upon him, and he expired placidly and calmly at twelve at noon on the fifth day, from the time he had first complained; which happened one day before I had visited him. It is a curious coincidence, that Lieut. B——, of the same regiment, who was taken ill at nearly the same hour, died also this day, within half an hour of Capt. I——.

    Here was a case without any remarkable symptoms of determination, except towards the head, and I am of opinion, it must have been of the serous kind, as the face never appeared

peared flushed or red, which would have probably happened, if the vessels of the head were surcharged with blood. I therefore judged purging the most proper evacuation; though I regretted afterwards, that I had not bled him, and that I had not insisted on applying blisters early. But we always regret where we are unsuccessful. Another part of my plan was to excite perspiration, and alter, if possible, the given state of the body; when these views failed, I had only to support the constitution and vital powers by such means as are commonly used on similar occasions. In cases like the present, where the vital powers are gradually sinking into a hopeless inactivity; where the pulse is hardly felt; where the patient does not complain perhaps of any pain; I have sometimes used the warmest and most stimulant powders, with some effect. This hint I took from the practice of my learned colleague Dr. JACKSON. I have prescribed brandy with Jamaica pepper, and given pills made up of Cayenne pepper, camphor, and opium. What will not one try, that can afford the smallest chance, in such a miserable dilemma, when he sees the common efforts feeble and useless? By these means, I was once successful, in the case of

a soldier

a foldier in the 18th light dragoons, of whom I had abfolutely defpaired; the pulfe was nearly gone, convulfive twitches, were every where felt; his extremities were cold, and he exhibited every appearance of diffolution. He took in the courfe of fix hours, 20 grains of Cayenne pepper, fix of camphor, and two of opium. The pulfe gained ftrength, the extremities became warm, and the features, which had fhrunk, and affumed the afpect of death, began to fill out and have expreffion; the anguifh in the eye vanifhed, and in the evening he was able to articulate. This was a kind of refurrection from death. One inftance of this fort juftifies innovation, and occafional bold practice. His name is FITZ-GERALD; and I believe he is now alive with his regiment. When common practice fails, and common expedients; it is a facred duty to vary our means, and mode of treatment; otherwife we become the idle, inhuman fpectators of death. It is no argument, to fay, that where we have no leading and precife principle to guide us, there muft be danger, and we muft practife in the dark. In our moft decifive practice, there is a great deal we do not accurately know, and were we only to act, where we have thefe precife principles,

ciples, the cafe in general would be committed to nature. If we affume the name of phyfician, let us merit the appellation, and give affiftance. We muft either act boldly or do nothing. Practice in my opinion, has hitherto been too tame and feeble; too much has been facrificed to prejudice; and a veneration for opinion. We neither think or act for ourfelves in medicine.—Fear has depreffed us; and we have furrendered our reafon to fyftem and doctrine. Unfuccefsful cafes, carry perhaps more inftruction, than the fuccefsful. —We are never to defpair; it is a common, but a good medical maxim, to guide the phyfician, that while there is life, there is hope; we ought never to ceafe to act, whilft the body can be acted on. While the vital fpark animates the frame, though it may be dim and feeble, it may be rekindled, and ftrengthened; the fibre may again produce the animal phænomena, in their common order; and the phyfician enjoy the unequalled happinefs of recalling a fellow being to exiftence. But we are more particularly called on for exertion, in a fever like the Remittent of St. Domingo, which has hitherto baffled every attempt, and committed unheard of deftruction; we will be juftified in every expedient, and humanity itfelf will

L apologife

apologife for boldnefs and innovation. I am not without hope, that we fhall yet cure this formidable difeafe.

We now come to confider the effects of Warm and Cold Bathing in the Remittent. Warm baths have been ufed in medicine from a very early period; they were recommended by the Greek and Roman phyficians in the cure of many difeafes. Among the Greek phyficians, there is reafon to believe that cold bathing was practifed with great boldnefs; though they are deficient in detail. In warm climates, tepid baths have always ranked among the luxuries of the great and voluptuous. But independent of the pleafing fenfations created by them, they become particularly effential from other reafons. The matter of perfpiration incrufts and refts on the fkin, fo as to form obftructions, and in fome degree block up the exhaling veffels. The tepid bath, by wafhing away thefe impurities, and ftimulating the veffels, enable them to carry on their functions with eafe and advantage. It is inconceivable, the hilarity and pleafant feeling, which the warm bath produces. The fpirits are enlivened; and a cheerfulnefs enfues; as if a burden was removed;

removed; as it is often expressed. But besides these evident uses, of tepid baths, in relaxing, cleansing, and stimulating the cutaneous vessels; there is another purpose of great utility. The warm bath, always increases sensibility, and leaves the body in a situation to be acted on. Sensibility is the great characteristic of animated matter, it is what chiefly distinguishes it; and forms the great basis on which changes are produced. Without sensibility, we in vain attempt to act on the human system. In this view I have often used the warm bath, and have found it highly beneficial. Cold bathing also has numerous advantages in warm climates and their diseases; it renovates the strength of the animal fibre; and by the sudden shock agitates all the vascular system, producing in it, a very quick change. But its chief effects are, to impart tone and strength to the vessels, to invigorate the general system, and by the agitation, remove obstructions in the more minute ramifications, of arteries and veins. In this view, I have found cold bathing very useful. Cold bathing has been of late years introduced into the treatment of fever. The Indians of America, followed this plan very succesfully, and in a very curious manner.

They strictly pursued the plan of effecting a total change in the system. Their manner was, to shut the sick up, in what is termed a wigwam or hut. They took a large stone, and heated it in the fire till it became red; it was then taken out and a bucket of water poured over it; the smoke or warm vapour could not escape; so that the patient was not only breathing this warm and humid atmosphere, but was, as it were immersed in it. By this treatment, a perspiration was usually brought forth, in the very height of which, the patient was carried out and plunged into cold water. This practice, which we should consider very hazardous, is said to succeed wonderfully amongst them, and to banish fever in most instances. We do not exactly know the species of fever which reigns amongst them, but from what I could learn, it appears to be of the Remittent form. At any rate, the fact of their effecting cures, in this manner, in the early part of the disease, is perfectly ascertained. It is a daring, bold practice, but if success crowns it, why not adopt it.

I remember when in the East Indies; on board the Airly Castle Indiaman, some cases of of Remittent occurred at Diamond Harbour, where the company's ships anchor. Several perished in spite of every attention; one of them however in a fit of delirium jumped out at one of the ports. He was immediately picked up, rubbed dry, and put to bed. His senses returned instantly, his pulse became more regular; he fell into a profound sleep; and next morning there was a complete remission. He recovered afterwards very speedily. I recollect another instance perfectly similar, on board the Princess Amelia East Indiaman, in the same place. The Remittent had carried off more than half the ship's company; though every assistance, every comfort the sick could have, was procured by Capt. Millet, the humane commander of the ship. A seaman of the name of Davies, a very stout athletic man, in whom the Remittent had at times alternated with epilepsy, jumped overboard; at the moment he did this, an alligator was along side the ship. He seemed to become at once sensible of his danger, and swam with great vigour till he was assisted. I saw him the moment he came on deck; his countenance, which before was grim and unpromising, assumed a more mild and temperate aspect; his pulse, which had been extremely quick and feeble,

feeble, was now flower and fuller; and his recollection, which had been confused and indistinct, became clear and accurate. I directed him to be washed over with brandy, and put to bed; he fell into a profound sleep, which terminated in an universal perspiration, warm, and profuse. The consequence was a very distinct remission, and a speedy recovery.

These cases made a strong impression on my mind, and I was determined to take an opportunity of imitating a practice, which accident had pointed out. I had not then seen the book of my friend Dr. JACKSON. An opportunity was soon furnished, of trying it further; upwards of thirty men were in my ward, at the Diamond Harbour hospital, and I commenced dashing buckets of water over them from some height; but whether the water was not sufficiently cool, or the patients being in expectation did not feel the shock, or that the circumstances were really different; I do not know, but I was not by any means so successful, as I had hoped from the two cases, I have just related. From the time I perused Dr. JACKSON's book, I became more fully persuaded that cold bathing, or rather the dashing of cold water might be made very useful

useful in the treatment of fever. And accordingly, I sedulously applied myself to observe its effects. The beneficial consequences from cold water, seem to me to arise entirely from a revolution it produces in the given state of the body; by which the whole morbid phænomena are changed. In the very early stage of fever, before it has established its peculiar mode of action; before the re-action begins, I think the practice of dashing cold water on the patient, may be very useful. But after the fever has established its peculiar morbid action, after the circulation and vessels re-act, after determinations to particular organs have begun, I hold this practice less certain. Because it is not likely to banish the mode of acting then fixed; and the sudden energy of the vessels from so powerful a stimulus, may assist determinations, and promote an inflammatory disposition. Previous evacuation may perhaps guard against these mischiefs. In cases, where sensibility is much impaired, where the recollection is confused, where the system is as it were oppressed, and wants energy to remove the oppression, where the pulse is feeble and frequent, in such cases, I hold the dashing of cold water, to be one of the best and most powerful remedies. The general effects I observed

152   TREATMENT OF THE   [Chap. II.

served to result from this practice, where it was happily applied, were, an improved recollection, more cheerfulness of aspect; a diminution of heat and anxiety, the pulse rendered more full and equable; a tendency to sleep and perspiration, and sometimes a distinct remission. I have said that these were the general effects, where dashing of cold water on the patient was happily applied. It must not be concealed, that I have used it often without success, in cases, where I promised myself much from its use. I have not at times been able to observe, that it produced any great effect. We have yet much to learn from experience, on this subject. The duration of applying water, the repetition of it, and the peculiar circumstances in which it is most useful, are not as yet accurately ascertained. I am of opinion with Rush, that it will be most usefully employed, in cases, where there is much diminution of nervous energy. In order to imitate the Indian scheme, as nearly as I could, I often premised the warm bath, and whilst the patient was sitting in it, I had two or three buckets of cold water suddenly dashed on him. I employed the warm bath in such cases, merely to create a greater degree of sensibility, that the cold water might be

more

more acutely felt, and produce its change or action, with more certainty. Surprife adds greatly to the power of thefe remedies; we may often fail, becaufe the patients have fummoned refolution to bear the fhock. It is often impoffible to take them by furprife; when once your practice is known, they expect the cold water, whenever the warm bath is ordered. Befides, it is doubtful whether dafhing cold water on the body produces, the full effect of a plunge, by which every point in the fyftem is at once affected; and in the two cafes I have related, the cold was longer applied, than it ever is in dafhing. Thefe are points, for experience to determine.

In one cafe, where I perfevered, and repeated the application of cold water, I was fo fortunate as to fucceed; and to preferve to his king and country a very valuable officer, in moft perilous circumftances. I allude to Col. H—— of the twenty-ninth light dragoons. His dangerous fituation is well known to many officers at Port-au-Prince. Col. H—— is of a thin, fpare habit, but active, and rather vigorous. Soon after he arrived at Port-au-Prince, he was feized by the Remittent. From the very beginning, the vital energies,

energies, were remarkably overpowered; a delirium commenced with the fever; the pulse was feeble and very quick, the skin dry and locked, and the countenance expressive of anguish and danger. I removed him to a situation, where I could daily visit him as often as his case required. His bowels were emptied, and I directed his servant to expose some buckets of water to a stream of air, in a situation, where the sun could have no access, so as to have it as cold as possible; I then had him brought out into the viranda, and seated on a chair, at a time he was so feeble, that he trembled all over, and manifested a disposition to faint. When seated on the chair, I poured over him a bucket of the water, from the elevation of another chair placed behind him. He was then rubbed dry and put to bed; the consequence was, that his pulse became fuller and stronger, his recollection more clear, and his skin relaxed, with a gentle perspiration over it. This state continued till towards evening, when symptoms of a fresh exacerbation appeared. He was again taken to the gallery, and another application of cold water made as before, with the same effects. The next day, the water was applied three times; on the third some symptoms of determination

to

to the stomach came on, and some degree of coma; the yellowness too begun its appearance, and the pulse became fluttering, quick, and feeble. A large blister was placed over the stomach, and one on each ancle; the cold water was again repeated three times; and at each time two buckets of it, instead of one, were thrown over him; the effects were remarkable; the pulse became instantly more regular, the *vis vitæ* was increased, and recollection became distinct. The blisters rose well, the vomiting ceased, and the danger seemed to be averted from the stomach. The application of the water was repeated in the same manner, the fourth day; the yellowness became deep, but a complete remission took place on the fifth; when the bark was administered to prevent any renewal of the paroxysm. This was one of the most dangerous cases of the Remittent, in which I ever saw a recovery happen. I advised Col. H——, to proceed to Jamaica, from which he was obliged afterwards to go for Europe. He was affected with dysenteric complaints, and did not regain his strength. It is with great pleasure I add, that he is now in England perfectly recovered. This was the most decisive case of the effects of cold water I ever met with. I certainly attribute to it,

the

the whole of the fuccefs. Becaufe good effects fo inftantaneoufly followed the application; and I am convinced the remiffion was obtained folely from this practice.

In the cafe too of my amiable and regretted friend, Major C——, of the 56th, I had obtained, by the fame means, a complete remiffion; though Dr. JACKSON, from the beginning augured danger. The fever had been abfent for two days before we arrived in Jamaica. The inhabitants of Kingfton, who ufed to receive lodgers, were impreffed with an idea of the Yellow Fever's being infectious, and would not admit any fick from St. Domingo. I went from houfe to houfe to procure accommodations for him, and at length, very late in the evening, I fucceeded; after much difficulty. During the day he fuffered the greateft anxiety, from an eagernefs to quit the veffel, and get on fhore; we landed about eight in the evening, and he was conveyed in a gentleman's curricle, to his chambers. At this time he had fufficient ftrength to walk, and was fomewhat cheerful. He had his feet bathed and went to bed. He paffed the night uneafy and reftlefs; he fighed frequently, and I forefaw, that the agitation and anxiety he had fuffered coming on fhore, would

recal

recal a fresh paroxysm of the Remittent. I called to my assistance Dr. GRANT, an eminent physician of Kingston, and an opening medicine was prescribed. About eleven however, in the forenoon, the paroxysm made a distinct, and formidable appearance; the vital powers, were oppressed and sunk at once; several blisters were applied, and cordial medicines; but all was in vain, the disease baffled every attempt, and this truly valuable man, perished about ten o'clock at night; whilst I was supporting him in bed. This case proves the danger of anxiety and fatigue in recalling the fever to new action; such returns are always dangerous, as they find the patient in a state of great debility.

In another case, Major C—k's, 69th regiment, whom I saw very early in the disease, so as to give the warm and cold bath every fair chance, I proved ultimately unsuccessful; for though at different times remissions were procured, and the strength and spirits seemed to improve, yet the fever still returned. But from the beginning, the deep sighing was present, a symptom I have ever seen formidable; he had at times such distinct remissions, that Dr. JACKSON, whose sagacity is
seldom

seldom disappointed, thought he had a fair chance of recovery. In this case there appeared some symptoms of determination towards the liver, and Dr. JACKSON recommended mercurial unction, joined with volatile liniment, and accordingly a considerable quantity was thrown in, but without effect. Calomel too had been very liberally employed, a combination of all the plans was adopted, and the dashing of water in all its forms had a fair, distinct trial. The fever, however, had its fatal termination on the ninth day. Major C—k had been much exhausted by service, he lost his arm in Corsica, and was otherwise much hurt, and had then recovered with difficulty. His habit, was what we call in the West Indies bilious, and he had always lived freely. I cannot help remarking here, what I have often witnessed in the fatal cases of this disease; an uncommon calm fortitude, which perfectly foreseeing death, talked of it with the most heroic indifference; whilst the mind was serene, acute, and firm. This I have met equally among the privates and officers; and though in the course of the disease, they have expressed some fear of the event, and seemed anxious, yet when it approached the close, they became at once dauntless

dauntless and indifferent. I have seen men calmly settling their affairs, after the circulation had ceased for hours, to be perceptible. From what this serenity of mind, so general in this disease, has its origin, it is impossible to determine. A few hours before Major C——k expired, I came into his room, he looked at me stedfastly; and stretching out his hand, in which no pulse was perceptible, he said, my time is at last come; my feelings tell me I must die in a very short time. I am obliged to you for your exertions; he called his servant, and asked if a little wine would injure him; I told him not; he desired him to fill out two glasses, of which he begged me to take one, and holding the other himself, he observed, that we had passed many happy hours together, and that he now addressed me for the last time. He drank a little of the wine, gave directions about his burial, shook me cordially by the hand, and turned round in his bed, where he very soon afterwards expired.

I attended another case, with my friend Dr. Fraser, of the 18th light dragoons, whose professional abilities are very well known. This was a case of an officer of great promise,

in

in whom all the army felt particular intereſt. In this inſtance, aſſiſtance was given from the very beginning, and Dr. FRASER judiciouſly combined every mode of treatment, that had the leaſt chance of ſucceſs. The daſhing of water hot and cold was tried ſedulouſly. Calomel had been given combined with purgatives; blood-letting had been freely uſed; and towards the cloſe, bliſters and cordials, and yet after all the vigilance, and care employed, the patient periſhed. The ſituation in which this officer caught the diſeaſe, muſt have given the higheſt activity to miaſmata. The regiment was ſtationed at a ruined plantation, ſurrounded on all ſides by a circle of marſhes, from which the cauſes of the Remittent muſt have been conſtantly iſſuing in their utmoſt vigour. The conſequence was, that numbers fell down, and Dr. FRASER aſſured me little impreſſion could be made on them, by any means, he could ſuggeſt. Some of them came into the general hoſpital; they for the moſt part periſhed; and we could boaſt of very little ſucceſs. But though this proves, that in certain circumſtances of diſeaſe, we are not always ſucceſsful, it does not forbid the hope, that we may yet become more fortunate, from further experience. I am ſtill perſuaded

suaded, that proper management may do a great deal, and a happy combination of treatment, effect many recoveries. In the efficacy of dashing cold water, I hold very considerable confidence, in changing and altering the given state of the body, and introducing a salutary revolution. In my own case, when threatened with the Remittent, and when in fact, a paroxysm had begun, I found the greatest benefit from vigorous treatment. My head ached severely, the pain in my loins and back were intense, and I felt an almost insurmountable inclination, to recline and slumber. An instantaneous determination had been made to the stomach, and the vomiting had begun: in this situation, I directed my servant to prepare the warm bath, into which I immediately plunged; when I had remained there, about five minutes, I ordered him to have in readiness, three buckets of cold water, drawn from a well adjoining the house, I then sat up in the bathing tub, and in that position the three were poured over me. I felt a considerable shock, and my headach increased to great violence; I arose and was rubbed down with a rough towel, and instantly put to bed. When I laid down, I took a scruple of James's powder. The vomiting had ceased on my getting

getting into the bath. My headach abated after laying down, but my fervant faid that I fpoke incoherently. I fell fhortly into a profound fleep, and a general perfpiration broke out all over my body; when I awaked, I found myfelf perfectly relieved from all unpleafant feelings, and complained only of a fenfe of wearinefs. Some inclination to vomit ftill remained, but it gradually difappeared; and in the courfe of the day I was enabled to begin the infufion of bark, and took two pills given me by Dr. Scot, containing fix grains of calomel. I had no returns, and in two days was enabled again to take charge of my ward in the General Hofpital. From all the fymptoms, I am fatisfied, that it was a real attack of the Remittent, and that the change brought on by the bath, procured a happy termination of the paroxyfm.

I have now finifhed the remarks I had to make on the various means ufed to combat the Remittent of St. Domingo; it would appear on the whole, that blood-letting in the early part of the difeafe, is one of the beft means we have yet adopted; whether by preparing the body for the action of other remedies,

dies, or in its own nature preventing dangerous confequences.

I fhall add a few words more on the ufe of opium, which in a variety of modes has been employed in the cure of the Remittent. In Intermittents, I have feen the ufe of it, attended with the beft effects, adminiftered in the cold ftage, in the manner of Dr. LIND. It certainly brought on the hot ftage, and conducted it to a happy termination, with great comfort to the patient's feelings. And I have once or twice feen, the return of the paroxyfm, entirely prevented by taking a large dofe of laudanum. But I muft, in the ampleft manner, join my teftimony to that of RUSH's, in faying, that I found the ufe of it in the early ftage of the Remittent, attended with bad confequences, even, when the pain and reftleffnefs of the patient called on humanity for any means of relief. The headach was always increafed, the fyftem in general rendered more irritable, the fkin more dry and parched, and an evident debility followed its operation. But towards the clofe of the fever, where there are fymptoms of remiffion, more freedom may be ufed, without any bad confequence. And it is a means of accele-
rating

rating recovery, by procuring sleep, and refreshing and invigorating the system. I shall now proceed to state the treatment I pursued in the Remittent, before I left St. Domingo, and the grounds on which I planned my practice.

The Author's Treatment of the Remittent, after considerable Experience.

WHENEVER I was called to visit a person attacked by the Remittent in the manner already described, if there was any inflammatory disposition, or that the patient was a stranger lately arrived, I instantly bled him in proportion to his strength and the urgency of the case; the quantity can only be ascertained by the circumstances then present, and cannot be regulated but at the patient's bed-side.—No directions can be given in words, that would apply to any number of cases; as minute occurrences often guide the physician. I am however of opinion, that much depends on the evacuation being liberal at first; if the symptoms do not change, and the pulse retains its vigour or increases in strength, the evacuation may be repeated next day, but not so freely as on the first. After the blood-letting the patient was ordered into the warm bath,

and

and whilst sitting there, half elevated out of the tub, three buckets of cold water were dashed over him; he was then taken out, and well rubbed with a rough dry cloth, and put to bed well covered; the room was chosen airy and open, and the bed placed in such a manner, that no direct draught of air played upon it. As soon as he was put in bed, an injection was administered, and eight or ten grains of calomel joined with a scruple of James's powder, were formed into pills, and one ordered every half hour till their effects were produced; the patient was permitted to drink freely of lemonade, beef tea, rice or barley water, tamarind water, orangeade, or any light drink that was pleasant to the taste. If the fever did not give way to this treatment the bath and cold water, were repeated again and again, till some impression was made in changing the given circumstances of the body. —The pills too were continued, till the bowels were evacuated, and a disposition to looseness was brought on; and the skin released from that hard crusty feel, so unpleasant to the touch. In general, where I could employ the baths, I directed them three times a day, and uniformly found that good effects resulted. When casual symptoms occurred, immediate attention was paid to them. Of all the means used to suppress the excessive vomiting and remove the irritation, I think large blisters early applied are the best; but I imagine we are in general too late in applying them, and permit the determinations to be formed before we attempt to counteract them, when they are too powerful to be removed, and have already produced their fatal tendency. I think in every case, where the slightest irritability appears, nay, where there is none, that a blister should be applied over the stomach, so as to prevent the determination to that important organ; for what is the pain or inconvenience of a blister, compared with the security that the application may afford. I would recommend then, and I actually prescribed a blister to the region of the stomach on the second day; this does not interrupt any part of the treatment. When the vomiting has once commenced, the patient should be directed to swallow as little as possible of any drink whatever, but to moisten the fauces and mouth often, to remove that dryness which conduces so much to the sensation of thirst. It is in vain to prescribe the mildest liquids, the irritability is inconceivable, whatever touches the inner stomach is sure to be rejected with violence;

violence; and every time the ftomach is thrown into thefe convulfive motions, the difeafe is ftrengthened, and the danger increafed. Whenever the naufea and pain appear, fomentations fhould be applied, and continued frequently, after the blifters are even placed, or rifen; very foft flannel may be employed for this purpofe, wrung out of hot water, or decoction of chamomile, of which fome entertain a high opinion. All medicines fhould be laid afide, during the height of the irritation; neither cordials or fedatives will anfwer the purpofe; I have never met with any medicine that would for any time remain in the ftomach. But above all, we are to refrain from the ufe of purgatives or antimonials, medicines which produce their effects by exerting their firft action on the fibres of the ftomach itfelf, efpecially calomel and jallap.— We are in this ftage to truft intirely to injections, and to repeat them often. Broth, and other nourifhing liquids may be thrown into the body in this manner, and the fæces may be removed, by adding irritation to the common emollient injection. If the blifters heal quickly, frefh ones muft be applied, and the difcharge fupported by iffue ointment. From this manner of ufing blifters, I have feen the moft beneficial effects refult, nor have I ufed any remedy with more satisfaction and success in removing dangerous symptoms.—I do not remember a case, where blisters failed in removing this most dangerous irritability of the stomach, where they were early employed and persisted in. I have also seen blisters singularly useful in the latter stages of the Remittent, when the spirits flagged, when there was a disposition to coma, and the pulse was low and fluttering, with that insensibility so often present with such symptoms. In these cases I have successfully applied blisters to the neck and shoulders, to the ancles and inside of the thighs, they were not large but made very strong so as to act; and I have seen cases where I could attribute recovery to them alone. One medicine I must mention, which I have used with good effect after the irritability of the stomach had somewhat abated; it was a solution of white vitriol in peppermint water, with the addition of a few drops of laudanum; I used the proportion of two scruples of the vitriol to six ounces of water and thirty drops of laudanum. Of this mixture I prescribed a table-spoonful every half-hour, till the symptoms disappeared. Dr. Jackson used at times portions of burnt alum with good effect. These must act by their astringent power, which is applied in a small bulk

without

without diftending the ftomach. The veffels in the inner coats, previoufly furcharged and dilated, are thus contracted and ftrengthened, the diftention which made them fo irritable is diminifhed, and they acquire fome portion of their former tone and feeling. It is in this manner only, I can account for the good effects of aftringents in this ftate of the ftomach.

In the progrefs of the Remittent, efpecially when remarkable debility occurred, I ftill perfifted in the ufe of the cold water, and generally found that the ftrength was repaired, the pulfe rendered more equable, and the recollection more diftinct and more decifive. During the recurrence of this debility, I ufed camphor joined with nitre, and fometimes James's powder, as I judge with good effect, in opening the fkin; and where there was any tendency to fubfultus, opium was added in confiderable dofes. If in fpite of thefe means the pulfe ftill continued to fink, and the vital energies to diminifh, I had recourfe to the warmeft ftimulants, fuch as æther, brandy, cayenne pepper, brandy baths, &c. It is then, of importance to maintain and fupport the living phænomena, to roufe the dying arteries, and to diffufe ftimulus, from the grand centre

centre the ſtomach. However theory may criticiſe ſuch practice, experience will juſtify it, as ſometimes ſuccesful, and ſucceſs is the beſt comment on any mode of treatment. When remiſſions were obtained, and the diſeaſe ſhowed a diſpoſition to yield, the infuſion of the bark and vitriolic acid were preſcribed, and continued during the tedious ſtage of convaleſcence, when the patient was apt to fall into a number of diſeaſes, ariſing from the previous derangement and debility of the ſyſtem.

It was in this general manner then I conducted the treatment of the Yellow Fever as it has been termed; varying my means, in many particulars as occaſion required. I might have perhaps furniſhed a number of caſes in detail; but my conſtant occupation did not permit me to take down theſe medical hiſtories at full length. I noted the general outlines, and progreſs, the remarkable occurrences, the general effects of various treatment; but this was all I could do. Occupied from morning to night in the hoſpital, or viſiting ſick officers, diſperſed over a large town, it was impoſſible to be very minute. Before I proceed to diſcuſs the views on which
I eſta-

I eftablifhed my practice, I fhall mention one caution in the ufe of blifters which may prevent much inconvenience to the patient; I mean the guarding the feat of them with the utmoft vigilance from the flies. The moment the fkin is removed, they croud upon it, and depofit their eggs, which become in this neft a race of maggots, and often form dangerous and deep ulcers, pouring out myriads of thefe difgufting animals. The patient often feels excruciating pain from their motion, and their efforts to feed on the animal fibre. They cannot be banifhed without much torture. They form finufes, into which they retire, and elude the forceps or probe. They refift mercury, fpirits, and the ftrong folution of corrofive fublimate, in all which I have tried the duration of their lives, which in thefe elements they fupported for hours. The fuccefsful and decifive remedy is the oil of turpentine, which never fails to kill and banifh them, when it has proper accefs to their habitations. This remedy was firft recommended to me by my friend Dr. WRIGHT of St. Domingo. It frequently gives intenfe pain, and almoft throws the patient into fits; fo that care ought to be employed in the beginning to prevent the flies from alighting on the fore.

I fhall

I shall now state the grounds on which my practice was founded.—In the first place I adopted no one particular remedy to which in all cases I invariably applied, without the assistance of others. It is evident, that as circumstances of disease vary in almost every individual, so must our means be varied also, if we wish to meet the disease and fairly combat it. On this principle the supporters of blood-letting, and the prescribers of calomel are equally wrong. No invariable mode of treatment can be adopted with success in any one disease. I accordingly adopted and blended all the systems of management which have been offered in the Remittent. I was certainly more successful, after I adopted blood-letting than before, and in many cases among the officers; where I was called early, had no reason to be dissatisfied with my labour. In the General Hospital, I could not by any means boast the same success; the cases which came under our inspection there, were of the worst description in themselves, and we seldom saw them in the early stage. The surgeons of regiments seldom ordered their patients to the General Hospital, until the case became very dangerous; in such circumstances many recoveries could not be expected, and accordingly the mortality was very great.

great.—But to return; I before stated, in the beginning of this work, that the first effects of the causes of the Remittent, were to form determinations of blood to various parts of the body, but more especially the internal organs: That these determinations consisted in a larger portion of the blood being directed to particular vessels, distending and stretching them, producing an increased morbid sensibility, and all the symptoms of acute inflammation. That the cause of these phænomena, was a diminution of strength, in some parts of the vascular system, by which the balance of circulation was destroyed; and that the danger of the disease, consisted chiefly in the strength of such determination, and the importance of the organ to which it was directed. That these determinations actually happen may be inferred from the phænomena of the disease itself. The irritability of the stomach, the astonishing impatience it manifests in rejecting solids or fluids, the convulsive agonies into which it is thrown, the pain to the touch, are strong proofs of this fact. But dissections prove beyond all contradiction, that effects very similar to those of inflammation, actually take place; the inner coats of the stomach are often found separated

from

from the reft, and floating loofe with the fecretions of that organ. The aftonifhing quantity of a ropy clear fluid often thrown up, when
the patient has had no drink, proves clearly
an increafed fecretion in the ftomach, which
could not happen, without an increafed action
in the veffels, and a larger quantity of blood
than the ufual proportion to that organ. That
determination happens to the head, is indi-
rectly proved, by the coma often prefent; by
the flufhing in the face, and the vifible action
of the arteries about the neck and temples;
and directly by diffection, which fhows actual
derangement, and marks of fulnefs in the
veffels. But what is more to the point, in
cafes, where the indirect proofs have occurred;
great fuffufions of a clear fluid have been found
in the brain. This clearly argues an increafed
action of thefe veffels, and an unufual fulnefs
in confequence. I know, it is difficult to
draw precife and juft conclufions from the
ftate of the brain after death; becaufe in the
ftruggles of dying, and the peculiar circumftances of refpiration, during thefe agonies,
almoft every one dies in a ftate of apoplexy.—
And hence, the veffels of the brain appear over-
diftended on diffection, though this diftention
might not exift till a few minutes before death;

† but

but where the symptoms already stated have appeared in the head, and where intense pain has been present; and diffection afterwards confirms, I think the conclusion may be fairly made, that there was a determination to the veffels of the brain. The liver too is very frequently attacked by determination. In almoft every diffection, that I have either feen or heard of; the liver has been found somewhat enlarged and tumid, and the gall bladder commonly diftended and full. In one inftance which occurred, when I was at Cape Nicholas Mole, in a perfon under the care of my friend Dr. FELLOWS; the liver contained an amazing impofthume full of pus. It had hollowed out, nearly half the hepatic fubftance; and the reft of it was uncommonly large, and tumid. I had feen this man a few days before he died, he complained of fome pain in that hypochondrium, and I fufpected that the liver was affected. Dr. FELLOWS, with a laudable induftry opened and examined him; diffection in a warm climate, is not the moft agreeable manner of inquiry. It appears then, from unqueftionable facts that determinations really happen, and that the greateft danger arifes from them. The danger of determination, would seem intimately connected with a certain

tain tone of the veffels, or what has been called an inflammatory diathefis, and this again to depend on the abfolute quantity and momentum of the circulating mafs. If this view be correct, and it is the only one in which circumftances lead me, to fee the difeafe, the propriety of an early and liberal blood-letting is at once eftablifhed.

Thefe are decided modes in which the proximate caufe operates, and blood-letting appears to me the beft and moft likely means to avert danger; but when the bias towards determination is completely formed, it is then matter of great difficulty, to prevent its going on. When the morbid action is once begun in confequence of determination, it is not eafy by any means to reftrain it; but to diminifh the bulk, and confequently the momentum of the circulating mafs, is the beft means we can employ to prevent it's fatal confequences. The veffels muft act vigoroufly and be in a ftate of diftention to produce the effects we obferve in the ftomach itfelf; the inner coats cannot be feparated without confiderable violence, nor the organization of the veffels and coats could not undergo fuch complete derangement,

without

without great morbid action. But bloodletting diminishes distension, tone, and vigour in the vessels; and therefore seems best calculated to prevent danger. And it is of the utmost importance, that the morbid action in the stomach, head, or liver, should not at all commence; as then, our blood-letting and other means may be too late. The effect of a stimulus, or rather the action it produces, may, and does in certain circumstances, continue, after the stimulus itself is removed. Hence it is of importance, to prevent the commencement of morbid action.

A fact occurs, in the administration of purgatives, which illustrates this reasoning; it often happens in the exhibition of salts, that they are instantly, rejected from the stomach; but the purgative effects are produced notwithstanding. That is; the stimulus imparted by the salts to the fibres of the stomach, had begun a certain action there, which had gone forward and continued after the salts themselves had been thrown up. It is the same in determinations, if once they have been formed, and that a peculiar action has been produced in the vessels; though the momentum and bulk, of the circulating mass, may be afterwards diminished;

diminished, and the danger of the determination perhaps lessened, yet some of its peculiar effects will go forward; and be at times hazardous. So that prevention, would seem a more secure ground of practice; and blood-letting appears to me, to be the chief and best means of effecting it. I trust it has appeared from the above reasoning, that it is essential to perform blood-letting very early; and as liberally as the circumstances of the patient will admit. It will appear also, that if the action of the vessels is not diminished, after the first blood-letting, that it will be necessary to repeat it; till that effect is really produced. Much of the future events of the disease must depend on an early blood-letting.— It is to be observed, that this doctrine, more especially applies to the case of strangers, newly arrived; and possessing like the English, the full, irritable, plethoric habit, on which the Remittent establishes its conquests. It is in such habits, that determinations, are apt to be produced. But they may occur in people, who have resided for a long time in the climate, whose vigour and strength have been diminished by perspiration, and the relaxation, heat always induces. In these cases,

blood-

blood-letting cannot be employed with the same freedom.

I omitted in my plan of treatment to mention Friction on the skin.—Whenever we are able to induce action in the vessels, the supply of blood to them is increased, or a determination to that particular set of vessels takes place. In this view Friction was recommended, which by having a chance of exciting into action the vessels of the skin, would divert the force of the circulation to the surface of the body. For it has been observed, that the determinations are apt to happen towards the internal organs. Friction therefore may divert the circulation to the surface, the spasm, which commonly takes place, may be thus removed, and the determinations to important organs prevented. The James's powder was prescribed also in this view, as I have often seen it operate on the skin; whilst the calomel cleared the bowels and removed fæculent matter, which might prove highly prejudicial to the system.

Both the calomel and James's powders are powerful alteratives, and may conduce to change the given state of the body; though

I am not clear, that smaller doses of calomel, would not succeed better, than large ones, as these become purgative. And purgatives certainly must in some degree favour determinations. The action of the vessels in the intestines once excited, solicits a greater flow of blood towards them; and purgatives operate by producing this action. Whether they compensate for this by removing, what would prove highly stimulant and dangerous, in the intestines, may be doubtful, as frequent injections might effect this end. Purgatives appear to me a dubious remedy; they certainly irritate, although they produce evacuation, yet in this way, they may contribute to diminish the bulk and momentum of the circulating mass. In Dr. Rush's practice considerable stress is laid on their carrying the calomel briskly through, and he seems to think that it contributed to the good effects of it, to be thus hurried through the intestines. To his authority, I pay great deference; but I cannot clearly conceive how this method could contribute to the good effects of calomel. I have already, pretty freely expressed my opinion of its use, as an alterative.

I come.

I come now to speak of the warm and cold bath, which I so freely employed; and here I must recur shortly to the doctrine of proximate causes, and our general knowledge of fever. It has been already said, that we do not know, precisely and definitely, the nature of proximate causes, and therefore, that practice by indication; is often mere amusement, a fiction, by which we deceive ourselves and our patient. Our knowledge of fever, in its intimate and necessary mode of existing; is not much more extensive.—Theories have risen after theories, and again sunk into oblivion; they are perused by the curious, as monuments of the difficulty of the subject; and the defect of knowledge. In these circumstances, the physician, untutored by any rational instructor, must search out principles for himself; and try some new path, by which he may prove more successful in his researches. On these grounds, as I could not prescribe by definite indication, in the Remittent, when the various means already enumerated, failed of success; I attempted to change at once the whole given, or existing circumstances of the system; so as to change the morbid phænomena; and by thus introducing a new order of things,

have

have a better chance of curing the difeafe. It is evident if the whole ftate of the body, undergoes a revolution; that the morbid caufe, cannot in a new condition of the body, produce the fame phænomena as before. But it may be afked, whether I can afcertain, that the new order of circumftances will be lefs dangerous, than the former which have been banifhed by my practice? To this I reply, that no new circumftances or change, can be more dangerous, than the ftate, we attempt to alter; and that the revolution in the fyftem affords fome chance, and therefore, that there is a preference due to it.— In cafes, where few efcape, and where certain fymptoms form prefages of death; it is furely the duty of the phyfician to vary his means, and not obftinately perfift in any one method, which has not been fuccefsful. Inftead of adhering to blood-letting or calomel, to bark, injections, diluents, or any one method, I took advantage of them all, and combined or feparated them, as occafion might require. From an attentive view of the difeafe and its fymptoms, I drew fome general conclufions, which had an influence on my practice; but I was often placed in a fituation, to abandon thefe conclufions, and attempt

tempt innovation. As the moſt powerful means of effecting a change in the given circumſtances, I uſed the cold bath. And I premiſed the warm, in order to create a high degree of ſenſibility, becauſe, without ſenſibility, we have no baſis to act on; nor can any change be actually produced. I had the ſatisfaction to ſee many caſes, in which I had reaſon to congratulate myſelf on adopting and purſuing theſe opinions, and if even our ſucceſs is confined to a few inſtances, there will be good grounds ſtill, for innovation.

I have now pretty fully explained my view of the diſeaſe, and the grounds of my practice; and I am not without hope, if it ever is my lot to be again placed, where it rages, that I ſhould be more ſucceſsful, than heretofore. It requires a long experience to form accurate concluſions. I have now only to obſerve, that, I confined the uſe of the bark to the convaleſcent period, when it ſeemed to prevent acceſſions, and to increaſe the ſtrength and appetite. A long train of diſeaſes often followed the Remittent, all intimately blended with debility. Obſtinate incurable diarrhœas frequently appeared, which reſiſted any thing I could ever try.

try. Malignant eruptions, ending often in foul ulcers, were sometimes the consequence of the fever; œdematous swellings, loss of appetite, great languor and debility, often remained for a long time, in spite of every remedy. It was always my opinion, that strength could not be recruited, in situations, productive of miasmata; and therefore, I uniformly recommended, as soon as the patient could bear it, a change of situation and climate. The state of our garrisons, did not sometimes admit of the absence of officers, who ought to have had the benefits of a change.— Many languished away life, in this unfortunate imprisonment.

It has been remarked, by many practitioners, that ulcers in the West Indies are much more obstinate than in Europe; and I believe there has been ground for the observation. But I must confess, that I have been surprised, by the success I have seen, both in wounds and ulcers at St. Domingo. Operations have been performed with more than European success, by Messrs. WARREN, MONTAGUE, and BUCKLE; and I have seen ulcers of the most malignant aspect cured, under the direction of the same gentlemen.

Having

Sect. IV.]   FEVER OF ST. DOMINGO.   185

Having now finished, what I had to say, on the Remittent, and its treatment, I shall offer a few words more, on the clafs to which I have referred it.   Dr. JACKSON, with whom it is not fafe to differ, and for whofe authority I have the greateft refpect, confiders the Yellow Fever as a diftinct difeafe, entirely different, from the Endemic Remittent; and for this opinion he offers the following reafons: " In our enquiries into the hiftory of the Yellow Fever (fays Dr. JACKSON) fome circumftances prefent themfelves to our obfervation, which are not a little curious. It has never been obferved, that a negroe, immediately from the coaft of Africa, has been attacked with this difeafe; neither have Creoles, who have lived conftantly in their native country, ever been known to fuffer from it; yet Creoles or Africans, who have travelled to Europe, or the higher latitudes of America, are not by any means exempted from it; when they return to the iflands of the Weft Indies. Europeans, males particularly, fuffer from it, foon after their arrival in tropical climates; yet after the natives of Europe, have remained for a year or two in thofe hot climates, efpecially after they have experienced the ordinary endemic

demic of the country; the appearance of the Yellow Fever, is obferved to be only a rare occurrence; but befides, that this difeafe feldom difcovers itfelf, among thofe people, who have lived any length of time in a tropical country; it has likewife fcarcely ever been known to attack the fame perfon twice, unlefs accidentally after his return from a colder region. The Remitting Fever on the contrary, does not ceafe to attack fuch as have refided, the greateft part of their life, in thofe climates, or who have lived, after the moft regular and abftemious manner; a fact, which feems to prove, that there actually exifts fome effential difference, between the two difeafes, or which fhows at leaft that the revolution of a feafon or two deftroys in the European conftitution, a certain aptitude or difpofition for the one difeafe, which it ftill retains for the other."—The above facts, which are ftrictly true, do not by any means, weaken my conclufions, or conftitute any real difference in the difeafe; on the contrary, they feem to ftrengthen my doctrine. Africans and Creoles, live chiefly on vegetables; they do not poffefs that conftitution or habit on which I allege the Remittent eftablifhes its conquefts. They have not the inflammatory

matory plethoric fulnefs on which the feverity and acceffion of the fever feem to depend; but when they travel to Europe, or the higher latitudes of America, they do acquire this conftitution, by a change in the manner of their living; by doing what Europeans and Americans do; in fhort, by living in the fame manner. They are then, on their return from Europe, liable to the fever; and fometimes to its worft form. But ftill the difeafe is lefs violent in general, among the Negroes and Creoles, than among Europeans or Americans. Becaufe in their habits, the inflammatory diathefis, does not exift in the fame degree. After the natives of Europe have remained, as Dr. JACKSON ftates, one or two years, in thofe hot climates; efpecially after undergoing the ufual endemic; the Yellow Fever is obferved to be a rare occurrence. That is, after the conftitution, by a refidence of one or two years, and by undergoing a mild difeafe, lofes the inflammatory difpofition; then, any fucceeding attack does not proceed fo far, as to induce the yellownefs, which I have noted, as marking a dangerous gradation, and the worft ftage of the Remittent. All the fecurity, which

which people derive from a long refidence in a tropical country, arifes from the gradual diminution of the inflammatory habit, which chiefly feems to produce the worft ftages and fymptoms of the Remittent. The fever, for the fame reafons, feldom returns twice; but this is only a very general obfervation, for wherever the plethoric ftate is produced in thofe, who have refided for years in thefe countries; the fever is apt to return with fatal violence. Of this I have feen many inftances. Captain I.———, of the 82d regiment, died of the fecond attack at the diftance of twelve months from the firft, from which he had happily recovered. Captain R——— died after he had refided twenty months at St. Domingo. Refidence only improves the chance of exiftence by diminifhing the inflammatory diathefis. " The Remitting fever on the contrary, (fays Dr. JACKSON) does not ceafe to attack thofe who have refided for years in the climate, and lived in the moft abftemious manner; and hence, there muft be a difference between the Endemic and Yellow Fever." Now all this reafoning amounts fimply to this, that though the Remittent attacks thofe, who may have refided

for

for some time in a tropical country; yet it does not commonly proceed to that violent ultimate stage of danger, the yellowness. Because the inflammatory diathesis has been destroyed by the relaxing powers of heat, and perspiration. But they are not wholly exempted, they are only subjected to a milder form of disease, from the antecedent circumstances, of the constitution. Dr. JACKSON has stated the facts with his usual correctness; but we differ in our conclusions; to his candour, I most willingly submit my inferences.

With respect to the variety of forms into which the Remittent is divided, I cannot do better than use the words of Dr. JACKSON. He has described in the following quotation all I have ever seen, viz. " A species of " disease, in which signs of putrefaction, are " evident at a very early stage; which is ge-" nerally rapid in its course, and which usu-" ally terminates in black vomiting. Yel-" lowness seldom or never fails to make its " appearance, in the present instance; and " perhaps it is only this form, which strictly " speaking can be called the Yellow Fever.

" Secondly,

"Secondly, into a form which either has no
"remiffions, or remiffions which are fcarcely
"perceptible; in which figns of nervous af-
"fections are more obvious, than fymptoms of
"putrefcency, and in which yellownefs and
"black vomiting are rare occurrences. Into
"another form in which regular paroxyfms
"and remiffions cannot be traced; but in
"which there are marks of violent irritation,
"and appearances, of inflammatory diathefis
"in the earlier ftage, which give way after
"a fhort continuance to figns of debi-
"lity and putrefcency; which yellownefs
"frequently fucceeds, or even fometimes the
"fo much dreaded vomiting of matter of a
"dark colour. The difeafe which I have
"divided in the above manner, in three dif-
"ftinct forms, appears to be in reality one
"and the fame. The difference of the fymp-
"toms probably arifes, from very trivial or
"very accidental caufes; it is a matter of
"great difficulty to difcriminate thofe figns,
"which are effential and neceffary to its exift-
"ence. It is in fome degree peculiar to
"ftrangers from colder regions, foon after
"their arrival in the Weft Indies, and may
"generally be diftinguifhed from the remit-
"ting

"ting endemic of the county, not only by the
"obscureness, or total want of paroxysms and
"remissions, but likewise by a certain expres-
"sion of the eye and countenance, with some-
"thing unusually disagreeable, in the feelings
"of which words convey only an imperfect
"idea."

CHAPTER III.

SECT. I.

REMITTENT OF ST. DOMINGO.

Means of Prevention—Changes in the System from Heat—Preparative Course for a Hot Climate.

HAVING finished, all I had to say, on the Remittent of St. Domingo, I proceed to a very important subject, the means of prevention. This will necessarily include some preparation for the climate, previous to landing; with the most likely means of avoiding disease, after the troops disembark; to which some Observations will be added on Diet, Situation, and Exercise.

Numerous directions have already been given in various books for the conduct of troops

troops on board ship; the best and most concise, that I have met with are contained in a pamphlet written by Mr. STUART, surgeon in the East India Company's service, and addressed to the Court of Directors. I shall not therefore enter into any minute details on this subject; but observe, that if my remark be correct, that the inflammatory diathesis, in any constitution, creates danger, our preparation at sea, must be directed to diminish this tendency. The great benefits of cleanliness, good air, and dryness, are known to every one who has passed any time on board ship. The effects of passing from a cold climate into a warm one, are sometimes very suddenly felt. Head-ach, nausea, an increase in the celerity and strength of the pulse, a discharge of bilious matter; argue some derangement in the state of the solids and fluids. The general effects of heat are produced in the human body. The solids and fluids suffer expansion, but not apparently in the same proportion; the fluids seem to be expanded before the rigid fibres of the solids sufficiently yield; this may be inferred from the hæmorrhage, which often happens from the nose, from the feverishness and tension of the pulse, from the scanty perspiration, which occurs on our entry

O into

into warm climates. These would seem to argue, that the fluids suffering a sudden expansion, burst the barrier of the blood-vessels, before they had time to accommodate themselves, to the new bulk of their contents. Heat too, renders all the vessels more irritable, and appears to communicate a stimulus to the whole system. Hence the secretions in general are increased, except such as mutually supply the place of each other; as the perspiration, and discharge of urine are known to do. Perspiration relieves the system in two ways, first by diminishing the absolute bulk of the mass of fluids, and thus accommodating them to the solids; and secondly, by conducting off the excess of positive heat, in the process of evaporation. These objects are of the utmost consequence in the animal œconomy. Our great aim then must be to put the body in a condition, on our approach to a warm climate, not to suffer from the unavoidable expansion and change that must go forward. This is to be effected by diminishing the fluids, and lessening the irritability of the system. On our getting into the warm latitudes, before the heat becomes intense; we must begin our preventive means. All those who are vigorous, plethoric, or irritable in their constitutions,

tions, ought to be bled, in proportion to their strength; it is impossible to lay down precise rules. For this purpose, the transports, or ships of war, ought to lay to; that the motion of the ship might not create present or future inconvenience to those who are bled. After this general blood-letting, a dose of salts ought to be administered to all those, in whom the operation was performed. They should be afterwards, every other day made to plunge in the salt water, for which purpose, large tubs might be placed on the forecastle. This would cleanse the skin, and preserve the proper tone of the vessels from undergoing too sudden a relaxation. The troops at the same time ought to be put on a lower diet. The quantity of salt provision must be lessened, and if they have been hitherto, accustomed to ardent spirits, they ought now to leave them entirely off; and to substitute the less pernicious beverage of spruce beer, porter, or wine and water. But these drinks ought to be of the first quality, and approved of, by a mixed board of military and medical officers. Unfortunately the liberal supplies of government to the army, fall too frequently a prey to commissaries and contractors; whilst the officers and soldiers, are robbed of their just allowances.

lowances. Those who can live on a more liberal scale, and can command vegetables and fresh animal food, ought to diminish the quantity of the latter, as well as their former quantity of wine. For though the waste by perspiration requires, that the circulating mass be recruited, yet it is not necessary, that the whole of this supply should be in wine. The great basis of all our drinks should be water, blended with such portions of other nourishing fluids, as will not permit it to weaken too much, which perhaps it might do, were it not for this addition.

Before the troops are landed the same means already recommended ought again to be repeated; so that the inflammatory diathesis would in a great measure be subdued. Ardent spirits though they create a temporary strength and excitement, yet dissipate the strength more than any other means. The languor and debility of a debauch last much longer than the joyous moments which produced them. In this manner by frequent repetitions, the animal powers are destroyed beyond recovery; and many men, formed to delight society, become humiliating monuments of debauchery. The stimuli of ardent spirits, wine,

wine, and animal food, exhauſt the ſyſtem more than any others, and waſte its excitability more completely. Men capable of great exertions have almoſt always, been ſober abſtemious men. Walking STEWART, ſeldom taſted animal food, or indulged in wine, yet he walked over an aſtoniſhing extent of country, without hurting his conſtitution; and with more perſeverance than is commonly found. Dr. JACKSON, who follows the ſame plan, and has always led an abſtemious life, gives, in his own perſon, remarkable proofs of vigour, under this regimen; and though now paſt forty years of age, is more active, and more capable of undergoing fatigue, than moſt of our young men of twenty. At Port-au-Prince he ſeldom mounted a horſe, and yet he viſited every barrack, every regimental hoſpital, and every ward in the general hoſpital, ſometimes twice a day. And he did not ſeem fatigued in the evening. Independent of theſe direct inſtances of the benefits of a moderate abſtemious life, we remark that the natives of warm climates in general, unleſs corrupted by Europeans, adopt this plan of abſtinence. Among moſt of the Eaſtern nations, this ſobriety of life is enforced by the principles of their religion. Their wiſe legiſlators, legiflators, enfured their obedience by facred ties; by which they performed duties effential to their exiftence and happinefs; with more pleafure and fecurity. The Mahometans, though they do not abfolutely decline the pleafures of wine, referve the full enjoyment of it to their celeftial manfions. The Gentoos, find a fufficient reward on earth, and practife moderation without a bribe. Temperance, like other virtues, is its own reward. But whilft I recommend temperance and moderation, I do not mean to infinuate, that we fhould wholly abftain from wine or animal food; on this fubject I fhall be more explicit in its place. It requires prudence and care to relinquifh habits to which we have been long enured. It muft be done by flow and cautious degrees, or we run a great rifque in the attempt to improve. It would be rafh to reduce a man, accuftomed to drink a bottle of wine after dinner, to two glaffes; fuch a reduction might be really dangerous. Nature, and the example of the natives of warm climates, would feem to point out, that the fame diet which is neceffary in cold countries, is not fuitable to tropical climates.

Nature

Nature exhibits, in the tropical climate, the human fyftem, relaxed and debilitated, and without the fame powers and vigour which mark the robuft inhabitant of Europe. The example of the natives, founded on this difference, teaches a mode of living proportioned to the vigour of their animal powers. Making this the rule of our diet, it would be certainly proper to alter our manner of living, on getting into the warmer latitudes. If it be the effect of heat to expand and relax, and by fuch change to debilitate the animal fibre; it muft affect every part of the fyftem, and diminifh the vigour of every organ, which compofes the living body. Among the other organs, the ftomach muft feel this general influence; and in fact we find its powers diminifhed; the appetite for animal food is languid, and when the ftomach happens to be full of any thing that requires energy to fubdue it, there is a remarkable oppreffion induced; and other fymptoms of dyfpepfia. The nature of the food muft be proportioned to the vigour of the ftomach. If this is a true axiom, we muft conclude, that as the ftomach becomes weakened and relaxed, in common with other organs of the body; we ought to fupply it only with fuch nourifhment as it can readily manage and subdue. Animal food in any large proportion, requires considerable vigour in the stomach to digest and subdue it. The proportion of it which enters diet in Europe, should therefore be diminished in a tropical climate; and a preference given to vegetables. Light soups seem to me well calculated for the powers the stomach retains; they do not require any great vigour to subdue them, and appear with little change to be fit for immediate assimilation. The French have adopted this mode of living, and are more healthy than we are. They are remarkable for their light soups and wines, and the large portions of bread, and vegetables which compose their diet; all their plans and methods of life are directly calculated to diminish the inflammatory constitution on which the Remittent makes its most dangerous attack. But, independent of the ease with which digestion is performed, and the proportion established by this method, between the powers of the stomach and the resistance of the food; it possesses another great advantage; that it gradually lowers the habit to a secure and less hazardous standard. For should the Remittent make its attack, it finds the habit in a situation to make a proper resistance.

These

These remarks apply to the West Indies in general, but more especially where Remittents prevail.

It belongs to this section to remark, that much depends on the period of arrival in the West Indies; but this is difficult to arrange, with any certainty, the attempt may be made, but the elements controul our arrangements. When however Government can attend to a certain season, the troops for West India service should be embarked in September. They will then, in all probability arrive in November when the healthy period commences, and they have before them four months of a milder temperature, during which they can be seasoned without danger. This is the only part of the year for activity or exertion; it is the only time in which European soldiers can be useful. They may be exercised with safety and trained to fatigue. It would therefore in every view be a desireable period for embarking West India troops. This attempt was made in 1795 from this country, but the event proved highly disastrous and dangerous. However such a boisterous season is not a common occurrence, and we may still hope for better success.

SECT. II.

Method of treating Troops after landing—Situations to be chosen for their Residence—Manner of Exercise recommended—Different Posts examined.

HAVING made these general remarks on Diet, which, so far as the rules can be complied with, are applicable at sea; I shall now suppose the troops disembarked. The remarks which will be offered on this subject, will apply in general to the West Indies; I shall afterwards point out more particularly, what applies to the island of St. Domingo.

When troops are landed, the first object of the officers attention, should be to secure for them dry and comfortable quarters; and to prevent as much as possible, with extraordinary strictness, their having intercourse with the troops already in garrison. The effects of this intercourse are, commonly, riot, intemperance, and drunkenness. Instead of allowing them to run about the streets, and fatigue themselves with novelty; they ought to be restrained from any unnecessary exercise, with the

the moſt vigilant caution. The moſt poſitive and ſtrict orders on this head, muſt be iſſued, and enforced. During this period of confinement, which ought to laſt two or three days, they ſhould all have a doſe of cooling phyſic, ſuch as ſalts and manna; ſoluble tartar and jalap; with a variety of other preparations ſuitable to the purpoſe. They ought to be fed on ſoups, with very little animal food. All ardent ſpirits to be abſolutely forbidden. After undergoing this preparatory regimen, they may be taken out to exerciſe in the morning, with ſome ſafety. The troops then, on the following day ought to be marched, to a known healthy ſituation, well ſheltered from unfriendly winds, lofty, and dry. This removal never ſhould be neglected; for it is proved, by accurate obſervation, that the miaſmata, which produce the Remittent, generally require ten or fifteen days to produce their effect; or more accurately ſpeaking require an expoſure to them, of that duration, before the body is ſaturated and yields to their influence. This points out an abſolute neceſſity for changing the ſituation of the troops, as ſoon as poſſible, after landing, and refreſhing themſelves. For all the places or towns in the Weſt Indies, where troops are commonly

commonly landed, have been built for the purposes of commerce, with very little consideration of their healthfulness. They are in general on low grounds, and these grounds being situated at the foot of high mountains, are somewhat marshy, and therefore not fit situations for troops. To this Port Royal in Jamaica forms an exception, being a sandy dry soil, and reckoned in that island very healthy; though not in the same degree as Stoney Hill barracks, an elevated, well sheltered, and dry situation.

When the troops are thus removed, it is probable, that they will not suffer, in any great degree, from the Endemic of the country. For allowing, that the miasmata act upon them, the moment they land, it is not improbable, as they require a certain length of exposure, to produce their effect, that the troops may escape after three days delay. But granting that a certain portion of them is admitted; the change into another climate, with a purer air, may prevent the ill effects which might otherwise result. The miasmata do not seem to operate like the smallpox; and from the smallest conceivable particle, produce their effects, as decidedly, as if

a great

a great quantity was employed. For the violence of Remittents, which we suppose, arise from miasmata; seems to depend, on the quantity of them, applied to the body, where the attack takes place. Thus, at Port-au-Prince, where, there are large marshes, the fever attacks strangers much sooner, and proceeds with more violence, than at Jeremie; where there is not the same nursery for miasmata as at Port-au-Prince. And at Bizoton, which rises from the middle of a marsh, the Remittent attacks after a shorter residence, than at Port-au-Prince; because, there are larger portions of miasmata produced and applied to the system. It is therefore probable that a certain quantity of miasmata is necessary to produce the Remittent, and a certain length of exposure requisite to saturate the body. These are sufficient reasons, for removing new-landed troops, to well known healthy situations. When they are removed; their diet should be as much as possible made up with wholesome vegetables; and great care taken, that no excess be committed with fruits and acids, which bring on troublesome diarrhœa and dangerous cholic. Spirits ought not to be allowed; but spruce beer, or what is perhaps preferable, good porter may be

given

given without hazard. Each soldier might confume three pints a day; with advantage. If fpirits are ever granted, water fhould be added, to dilute them; perhaps half a wine glafs of good brandy undiluted, might be allowed after dinner. Soldiers cannot believe their exiftence fecure without ardent fpirits. And the officer may find it neceffary fometimes to concede fomething even to their caprice.

I now fuppofe the troops in healthy, well fituated cantonments, where they are to be trained for the fervice of a hot climate, with all the fecurity againft the invafion of the Remittent, that they can well poffefs. If they are in the neighbourhood of running water, or any convenient fituation for bathing, I would ftrongly recommend the cold bath, every morning, or every other morning. Unlefs the furgeon of the corps, points out unfit fubjects for this healthy exercife, they fhould be all ordered to bathe. An officer and the furgeon ought to fuperintend this operation, that no riot or play may take place, which might keep them too long naked, or in the water. When they have plunged and wafhed themfelves, they are to be rubbed dry with a coarfe towel, with which each of them
ought

ought to be furnished; and then retire to their barracks with a brisk pace. They will feel themselves light, invigorated, and cheerful. They resist in this manner, the relaxing effects of heat, they acquire strength without the inflammatory habit, and they become fit for service, without fearing disease.—This, as Thompson has expressed it, is the purest exercise of health; the kind refresher of the summer heats. And the same poet adds, 'that the Roman arms, which subdued the world, first learned to subdue the wave.' At the mess of soldiers an officer ought always to be present, to regulate their conduct, and see, that no impropriety is committed. This is particularly necessary in the West Indies, where all the caution and vigilance employed is hardly sufficient, to prevent the men from getting a poisonous kind of rum, which destroys numbers. I saw at St. Domingo a striking example of the benefit of this attention, in the royal, or first regiment. Colonel Green, with the most laudable and indefatigable zeal, was never absent from the mess of his soldiers; he regulated, ordered, and conducted every thing. The effects of this management were astonishing; the royal were the most orderly, the neatest, and best looking

men

men in the garrison; their deportment and appearance distinguished them at a distance; and other regiments looked up to them as objects of imitation. This sort of attention in officers, greatly improves the military ardour of soldiers, they feel their own importance, they are sensible of the kindness shewn them by their superiors; and they will not disgrace their friendship. At all events this vigilance must prevent riot and drinking, or the smuggling of rum into the mess; for which the most ingenious stratagems are contrived.

After regulating situation and diet, the next thing of importance, is military exercise; which is to complete the soldier for the purposes of his profession. It has been a general practice among officers, and from the best motives, to prevent the men as much as possible, from running about in the sun, or being even exposed to it for any time. This restriction is certainly very proper for the first few days after their arrival, but no longer. The soldiers who arrive in the West Indies are destined to serve in a hot country. They must, as service may require it, be exposed to the action of the sun for many hours; and often for a whole day;

day; is it not abfurd then, not to initiate thefe men into a gradual tolerance of what at fome period they muft undergo? It is not from a dark chamber, we would bring a perfon, to fit him, to bear the fplendour of light, without hurting his eye. Nor would we train, in a heated apartment, a foldier for the cold region of Nova Zembla. Equally improper, is the method of chafing troops to their barracks, whenever they appear in the fun. The confequence is, that troops thus educated; in the cool fhade of their barracks, are rendered unfit for any fervice; in this retreat they languifh and fleep their time away, they acquire indolent habits, they become relaxed; and for want of better amufement get drunk. When the fervice requires them to quit the fhade, and march into the field; they are incapable of fatigue, the firft beam of the fun injures them; like tender plants they fhrink from the breeze, ficken and die. Nor is this furprifing; it is to them, a fudden change, they plunge at once into fevere exertion; without being gradually inured or trained to it; and they fuffer accordingly. It is aftonifhing that this prejudice fhould prevail, and be even encouraged by medical men, as I have frequently feen. If indeed foldiers could fight, with umbrellas over them, or command an eclipse, whilst they were in the field; the present method of training them, would be proper and useful. But as effeminacy has not yet corrupted, or miracles assisted our armies, we must recur to methods founded in common sense, and common reason. What is it, that gives the superiority to black troops over ours in a warm climate? Is it not the capability of enduring the heat of the sun without danger?—and ought it not to be our aim gradually to bring our own troops to their level? Nothing will do this but education; a gradual habit, which at last steels the constitution, and fits it for any exigency. I have seen at Port-au-Prince, many men drop down on short excursions, affected by the heat. If they had been gradually inured, this would not happen. I brought myself to bear the sun's utmost heat without any disagreeable sensation. My profession required, that I should be able to go out at every time of the day in which I might be called on; I never used an umbrella, and I rode all day, wherever my assistance was required without any inconvenience. But I acquired this habit gradually. Dr. JACKSON acquired the same habits, and walked daily in the sun, without suffering

suffering any injury.—With respect to troops, it is necessary to form this habit of enduring the sun very gradually. Let them begin to exercise at five o'clock in the morning; and continue this practice for two or three days, then come on, to half after five, to six, seven, or eight o'clock, prolonging their stay in the field each day; and making the hour progressive. After a habit of performing their exercise at eight o'clock, let them be gradually brought on to twelve in the forenoon, the warmest period of the day; and at first detained for short intervals, which may afterwards be increased to any time. Troops will thus become highly useful; and in time be equally fit with the natives of the country to undergo any fatigue or service. I appeal to common sense, on this question, to the experience of every one who has attentively viewed troops on any expedition in a warm climate. It must be evident, that it is proper to train them by degrees, to form and establish a habit of bearing heat, to which the nature of the service, must unavoidably expose them, some time or other.

When at Madrass, in India, I had an opportunity of observing, that the troops in that country

country were generally exercised, at an early hour in the morning without any variation; and that they were sent back to their barracks with a sedulous care, to guard them from the sun. I understood at the same time, from a number of officers, that on service, many of these troops suffered from the sun. Having had an opportunity of mentioning this subject to the late Sir ARCHIBALD CAMPBELL, then governor of that settlement; he was struck with some observations which arose in our conversation; and that excellent officer adopted the plan of progressive exercise, from an early to a more advanced hour of the day. At first, troops will not bear a long period in the heat of the sun, but by degrees, they may be so inured, as to undergo long marches, without any inconvenience. This is a subject of great importance; as a very different system is at present carried on in the West Indies; a system which unfits the soldier for exertion, and exposes him to new perils; against which we might easily guard, by a little attention.

I am so convinced, on the propriety of bringing up soldiers, to a gradual endurance of heat, that I am astonished, it should not have been long since

since adopted; but prejudices do not easily give way; and the sun has from immemorial time been reckoned unfriendly to the European constitution. The restriction is highly proper at first. It is a curious fact, proved almost by every one, that those who arrive from England in a warm climate, or from the latter in England; endure the heat or cold better, for the first season, than during any subsequent one. This is a dangerous prepossession, in either case. It is difficult to explain, how the sensation arises. It would seem to me, to depend on this principle chiefly; that we conceive the heat of the East or West Indies, greatly beyond its real standard. And the inhabitants of warm regions imagine the climates of Europe, to be much colder than they really are. When people arrive under this impression in these climates, they do not find the heat or cold correspond with the degrees, imagination had fixed, and therefore they brave all caution, until the fervour of fancy has abated, when they feel the heat or cold as they really are.

After the troops have been stationed for a few weeks in healthy situations, they might be permitted for a few days at a time to do duty in less salubrious places, and then, return again to their former stations. They would thus become gradually habituated, and at length resist the influence of the miasmata. The want of such situations at St. Domingo proved highly injurious to our troops. When the troops from Ireland arrived at the Mole, they were obliged to live for weeks on board the transports, where an infectious fever raged, and made great havoc amongst them. This would not have happened if they had been landed and encamped, on the neighbouring hills, or if there had been proper barracks to receive them; the consequence was, that numbers perished; and that little army, which originally consisted of five thousand men, was very shortly reduced to fifteen hundred. There were no places prepared for the sick, or their necessaries. Military and medical stores were landed in haste, and strewed the shores, like fragments of a wreck. The exertions of the medical gentlemen at this post were extraordinary. Mr. WEIR, the inspector-general of hospitals, took every possible means, of providing in the best manner for the sick. He took the duty of physician, and fatigued himself in every department that required his presence. Dr. JACKSON exerted his usual humanity;

humanity; and Drs. MASTER, CLEGHORN, and FELLOWS, had their ample fhare of employment. Yet from not feeking a more healthy fituation, the troops perifhed rapidly. In fact, they imported from Ireland an infectious fever, which, for a fhort period, raged, independent of the endemic of the ifland, and did great execution. In fuch circumftances, the exertions of medical men proved of no avail; terror fpread wide amongft us, and increafed the conquefts of death. Of fuch importance is it to chufe healthy fituations.

It is now time to fpeak of the healthy pofts at St. Domingo, which unfortunately are few in number, and therefore eafily defcribed. It may be the lot of Britifh troops in the courfe of war, to land again, in this unfortunate ifland. If this misfortune fhould ever happen, let us take fuch precautions as the nature of the country, in its prefent fituation, will admit.

The Mole, from its commodious harbour, is generally the place where troops are firft brought; but no delay ought to be made in this fituation. Such troops as are really neceffary for the defence of the poft, muft be landed, and after undergoing the preparatory regimen

regimen already mentioned, they are to encamp on the brows of the hills, beyond the town, in the manner which general WHYTE very judiciously adopted. They are there less exposed to danger, and form a cordon of defence round the garrison. It has been very clearly proved, that lofty situations are not by any means the most healthy in marshy countries; because they are more exposed to streams of miasmata, from the very circumstance of their elevation. We are therefore to chuse dry and well-sheltered situations, especially against the land winds. We are to take care, that no marshes of any extent are to windward of us; and that we have, if possible, streams of running water in our neighbourhood.

After landing at the Mole, what may be absolutely necessary for its defence—the rest of the troops, should proceed for Jeremie, a district of the *Grand Anse*; by far the most healthy situation in St. Domingo. It was to this place, that the convalescent French, used to fly as to Montpelier, for health; and generally succeeded. The inhabitants and troops of this quarter, wear an European aspect, when compared with the sallow complexion of their neighbours. The country is dry and lofty, streams

of

of water rush down from the mountains, and the sea breeze, cools and refreshes the air. Here then is the situation, where we ought to land the bulk of our troops for the service of St. Domingo. After they have been in this healthy situation for some weeks, they may be sent to St. Marc's, or the Mole, and last of all to Port-au-Prince. But during their residence at Jeremie, they ought not to be confined to the town; but be scattered over the most healthy plantations, which for this purpose, may be converted into useful barracks. To this the French proprietors cannot object, as they derive from the troops defence and security. The barracks at Jeremie were judiciously erected under Colonel MURRAY's inspection. If it becomes necessary to send troops, to that fatal spot, Port-au-Prince; they ought not for any time to remain in town, which in fact is a nursery of miasmata. They ought to be encamped on the brows, of the mountains which surround that town, in a curve line from Tourgeot to Bissoton. This plan of incampment, was pointed out by Dr. JACKSON, who travelled over the ground; but it was not adopted from some frivolous objection. It was alleged, that the inhabitants of the town, could not be trusted; or that they

they would lose confidence and fly the town. This could not well have taken place, when Fort Royal, formed one point of the Semecircle and Bissoton the other, whilst the connecting line was a chain of posts—however, the plan was not adopted.

The greatest attention ought to be paid, that troops are not placed, in situations known to be unhealthy. The spirits sink, and the operation of fear renders the access of disease more easy and certain. To places remarkably unhealthy, as Port-au-Prince, the best seasoned troops ought to be sent; and placed in Fort Royal, which commands the town so very completely, that in case of commotion, its artillery might very soon reduce it to ashes. The duty in the town might be performed by our best colonial troops, under the command of honourable and well-tried officers, such as the baron Montalembert or Desource. The duty of Bissoton must fall to them also, being the most unhealthy of all our posts. The colonial troops would exist there, though it be certain death to ours. The British in this arrangement, are to occupy the brows and declivities of the mountains; dry and well sheltered posts. They form a grand outline, and defend the town, at least much more effectually,

effectually, than if they were included in its hospitals. They would be thus removed, from the great source of disease, the miasmata; and they would occupy a country where they would find a purer atmosphere, more chearful prospects, and more healthy amusement.

The Croix de Bouquet is found tolerably healthy, by the French inhabitants, but it has proved fatal to the few British who resided there.

The plantations spread over the plain of the Cul de Sac, form the worst possible barracks, because the whole of that wonderful and fertile spot, is itself a marsh, where a constant exhalation is going forward; and the accommodations for the officers and soldiers are of the worst description; from the devastation of the unhappy negroes, who are willing to erase every monument of human industry, and every trace of their former labours. It is indeed melancholy, to ride among these wide-extended ruins. Every where, marks of opulence, elegance, and commerce, all now levelled with the ground, by many of the hands which assisted to rear and protect them.

Our

Our troops have also suffered very much at L'Arcahaye, which formerly was reckoned a very healthy situation. Whether it was, that LE POINT the French commandant, did not chuse any interference, from English officers, or that he really had no good situation for them or their troops; it is certain that all the officers sent to that quarter made complaints of their treatment. This post therefore should be chiefly occupied by colonials.

St. Marc's, formerly pretty healthy, has proved very fatal to our troops. Part of the ninety-sixth regiment landed there, and were soon exterminated to a man. But I believe an infectious jail fever, had raged amongst them; ere they left the transports, and carried off great numbers after they landed. On a view of St. Marc's, it would appear to be advantageously situated for health. It is sheltered on the land side from the pestilential land winds, by very lofty mountains; the soil on which the town is placed appears dry and sandy, the skirts of it are washed by the sea, which carries off many impurities, and the sea breeze, blows with little interruption. Yet after all, this place has been very unhealthy to the British troops. I think the opening of a ditch round the

the town, which expofed a great furface for exhalation has greatly contributed to this unhealthfulnefs. Indeed the inhabitants themfelves made this remark, and feemed to date the commencement of ficknefs, and fever, from this period.

LEOGANE was alfo very unhealthy, when we had poffeffion of it; it is fituated in a marfhy plain. Nor were we much more fortunate in Mirebalais, where it was hoped we fhould have enjoyed much better health. When Dr. JACKSON vifited this poft, he ftrongly recommended, that the Britifh troops fhould be moved from the ground they then occupied, to a more healthy pofition, and in fact an order was given to move them; but the French commandant found means to evade it, and detained them; till they all nearly perifhed. He was engaged in fome lucrative contracts of fupply; and the Britifh were good cuftomers. The 82d regiment one of the fineft, I ever faw, was ftationed in this quarter, to which they were fent, as a refuge from the devaftation of Port-au-Prince, where numbers of them had already perifhed; it is melancholy to relate, that they here found no fanctuary, and returned in a

few

few months not twenty men strong. In September 1795, this regiment was reviewed by Sir ADAM WILLIAMSON, complete in all its officers, and men; nine hundred and fifty strong; in September 1796, they had not fifty men fit for duty, and in November, arrived at Port-au-Prince from Mirebalais, with hardly their complement of non-commissioned officers. This beautiful corps, in the space of less than a year, lost upwards of eight hundred men, and twenty officers. Such is the melancholy devastation of this climate.

All our troops then, ought, if possible, to be landed at Jeremie, and after undergoing there, a seasoning of some weeks, they may be distributed in rotation to other posts; going first to the least destructive, and continuing a change, till they are able to do duty at Port-au-Prince. The nature of service may occasionally oppose these arrangements, as we must be guided by circumstances, and the position of our enemy. But where these rules can be observed, I am convinced they are of importance, and might be regulated and observed without expence, and with real benefit to the service.

I omitted

I omitted to mention, that the Grand Bois, not very diftant from Port-au-Prince, is reckoned by the inhabitants equally healthy with Jeremie itfelf. To this place, troops newly arrived might be ordered. I confefs there are fome difficulties in the way of thefe arrangements from our uncertain poffeffion of the ifland, and the neceffity of avoiding expence, in a country which has already drained the Britifh treafury; as well as proved the grave of the Britifh army.

I remarked in a former part of this work, that a chief difficulty arofe in reftoring Europeans to health, from the continual application of miafmata, and the relaxing powers of heat itfelf. Once the patient is weak, we feldom fucceed in giving him ftrength. Languor and debility prevail, in fpite of every means we employ. The cold bath, and the ufe of bark, though they maintain the patient in ftatu quo, without any progrefs, yet feldom fucceed completely; becaufe they are powers not conftantly applied, or uniformly acting; whereas the powers of heat are conftantly and uniformly applied and acting; their effects then cannot be counterbalanced, by the action of temporary powers, which

are

are only present for a short period, and effect changes of very short duration. The action of climate, on the human body is perpetual, the action of medicines, temporary and short. The effects of climate then, must always prevail, over the action of medicines. Convinced of the truth of this doctrine, by fatal experience; I was always of opinion, that a well-chosen situation at Jeremie, for a convalescent post, would be of the greatest utility. Here the emaciated officer and soldier, placed in a different climate, and surrounded by new scenes, would be invigorated and recovered. The voyage itself would contribute to this desirable end; the movement, anxiety and novelty, would divert the mind from brooding over misfortune, and give to thought a new and more pleasing direction. The coolness of Jeremie, which approaches an European climate, would contribute remarkably to recovery. We are pleased in finding ourselves in situations similar to our native country; our habits are soothed, and our constitutions acknowledge a kindred sympathy; whilst our progress in strength every day delights us. From the first moment I ever saw Jeremie, I was of opinion, that it was a situation highly calculated for a convalescent valefcent hofpital, and poft; and that it might not only be itfelf guarded by convalefcents, but alfo send back to the other pofts feafoned men, who having undergone the endemic, and recovered, would feel more confidence and fecurity. In this idea, I was joined by my friend Dr. WRIGHT, with whom I have often coincided in medical opinions; and we jointly gave in a paper to Sir ADAM WILLIAMSON, recommending this meafure. He was however preparing to return to Europe, and did not chufe to enter upon a plan, which might be more productive of expence, than he could forefee at that period. The plan was accordingly abandoned. Mr. WEIR has fince adopted the fame opinion, and partly converted Jeremie into a convalefcent ftation. Dr. JACKSON, who vifited this quarter, intended to have fent there all the convalefcents of Port-au-Prince; but though he recommended it warmly, the fcheme was never fully adopted. There were numerous veffels employed by government, at a vaft expence, which often lay for a confiderable time idle, that might have thus been moft ufefully, and beneficially employed, for the benefit of the troops. Jeremie was a central point, to which the

Q   convalefcents

convalescents of the Mole, St. Marc's, Larcahaye, and Port-au-Prince, might have been easily sent, and from which they could again be returned with perfect facility. One of the government vessels, might have been fitted up as an hospital ship, with an assistant surgeon on board, and proper remedies, to convey the convalescents from the other posts to Jeremie. This ship, when not actually on service, might occasionally run to sea, with sick officers, and afford them a chance of recovery, which they could not have on shore. And if well armed, she might protect trade, and be a terror to gun boats and privateers. Such a scheme, though extremely simple, and necessary, was not adopted, after all the recommendation it received. Medical officers can never enforce schemes of health; they may recommend, but cannot execute. Perhaps on these points, their authority is too limited; though an extension of it might interfere with that absolute power, commanders must possess to ensure general obedience. The expence could not be great, when we consider the price of every soldier to government, before he is landed in that country. Many might have recovered at this post, who languished

languished life away, in the general and convalescent hospitals of Port-au-Prince; where, though they had good medical assistance, and as much attendance as the nature of circumstances permitted; yielded at length to the fatal and perpetual action of the climate.

It is true we had at Port-au-Prince, a building we called a Convalescent Hospital, placed conspicuously, in the most unhealthy part of the town. Elevated beyond all shelter, it stood exposed to every land breeze that blew, and arrested the floating miasmata, as they were blended with the air. In this hospital, it could not be expected, that many recoveries would happen; but as the soldiers sent there, had in general got over the danger of the first attack, they lingered for a longer period; and either died there, or returned again to the general hospital. The proportion of useful recoveries, was very slender. Indeed, the general hospital itself stood in no promising situation, it was low, and on the borders of a marsh. But on the whole, a better sheltered situation than the convalescent hospital.

Dr. JACKSON on his arrival at Port-au-Prince, surveyed minutely the state of the hospitals,

pitals, and was fenfible of their naked expofed fituation. He accordingly recommended loofe thin curtains to be fufpended before all the galleries, and to contain within their fhade, the whole ward, in which the fick were placed. This was of great ufe; it enlarged the bounds of each ward, by adding the gallery, which before, the fick could not occupy for the fun. It afforded them an agreeable change of place, and a cool retreat from the ward, which the breath of fo many fick, had rendered hot, and which was made ftill more diftreffing from the groans and complaints of the dying. But it had another advantage, by poffeffing a loofe texture, it admitted the air pretty freely; and when the land winds blew and were hot, they were rendered cool, by throwing water on the curtains. The hot winds are thus tempered in the Eaft Indies, and rendered pleafant and refrefhing; when they would be otherwife infufferable and dangerous. But befides thefe purpofes, the curtains might anfwer another very important ufe: They might probably purify the atmofphere, by arrefting the miafmata in their progrefs, and thus anfwer the purpofe of natural fhelter.

SECT. III.

*Confiderations on General and Regimental Hof-
pitals—Hofpital Corps, unfit for their Occupation—A Medical Board recommended with
large Armies—The Qualifications of Phyficians examined.*

IN this Section, I shall confider some circumftances, which could not enter so properly into the difcuffion of the fubjects we have been treating; but which are intimately connected with them. And firft, with refpect to General and Regimental Hofpitals.

General hofpitals have been reprobated by a number of officers, and phyficians, without being able to abolifh them. The eftablifhment of a general hofpital, is always an expenfive cumbrous inftitution. When there are a number of wards, the foldiers, who are able to move, vifit one another, and mingle into focieties and clubs, where military habits are loft, in riot or idlenefs. The attendants too, have a fimilar intercourfe, and as they muft have charge of the wine and liquors, prefcribed scribed for the sick; they are enabled to enliven their meetings, by a considerable share of gaiety. In this manner, the sick are often totally neglected, in spite of every vigilance on the part of those who attend them. Squabbles, noise, and riot, are the result of these associations. It has, I believe, been remarked, that though a soldier may enter a general hospital with all his military ardour thick upon him, that he will never return with it.—And I have seen enough to convince me, of the truth of the observation. Though the soldiers are in general very unwilling to enter, they are equally unwilling to come out; and linger their time in passive languor, or in the more destructive scenes of debauchery. The institution of general hospitals, throws on the medical staff, the whole duty of regimental surgeons; who are thus rendered idle and useless. It was thus at Port-au-Prince, previous to the arrival of Dr. JACKSON, who regulated the regimental hospitals on a better footing. There was indeed one belonging to the twenty-third regiment, which was very well managed; and with little expence to government, under the direction of Mr. BORLAND, surgeon of that corps. Mr. WEIR at the Mole, arranged the regimental hospitals in in such a manner, that they had very few indeed, in their general hospital. At Port-au-Prince, we seldom had less than two hundred, and often more. The reason was, that there were not, well endowed regimental hospitals, and that the surgeons, the moment a patient was reported to them, ordered him to the general hospital. The consequence was, the general hospital was crowded, and became the receptacle of all the army. Many arguments are in favour of regimental, in preference to general hospitals. The institution itself is conducted with little expence to the public; whereas a general hospital creates an enormous demand, and adds greatly to the expenditure of an army. In most regiments, there is a fund appropriated for this purpose; which if properly managed would prove in general equal to the expence. But if it should not, it would be much better for government to afford them a regulated supply under the inspection of a proper officer, than to institute a general hospital. A regimental surgeon must derive great assistance, from knowing the character and disposition of his patient; which in a general hospital cannot be so well known or ascertained. The patient is attended by his comrades, from whom he will receive more

tenderness

tenderneſs of attention, than can be hoped for in the indiſcriminate attendance of hired nurſes. Nor has the patient the ſame chance of departing from his military habits and becoming corrupted. The circle in which he is placed is narrower, and his opportunities to err fewer. He is placed more immediately under the eye of his officers, who, by frequent viſits, maintain good order in the ward. To theſe officers, they know they are to return when their health is reeſtabliſhed, and they are conſcious their behaviour will be noticed and recollected. In a place like St. Domingo, where numbers are at once taken ill, it may not be convenient to obtain houſes independent of public inſtitutions to contain all the ſick of a regiment. When this is the caſe, the general hoſpital, may be divided into departments for each regiment, and their ſurgeons in that ſituation, ought to attend them. If they require medicines or ſupplies, which the regimental cheſt, or regimental fund cannot ſuſtain, let them be ſupplied from the ſtores provided by government for the army at large. It may be aſked, how the medical ſtaff are to be employed in this arrangement, as the patients in the general

neral hofpital are recruited from the feveral regiments in the garrifon? The ftaff may be moft ufefully employed in fuperintending thefe hofpitals, in feeing the mode of practice, in correcting abufes; and in affifting, where their fervices may be moft requifite. This will afford the ftaff fufficient employment; with the occupation of attending the fick officers in their refpective quarters. In this manner, an amazing expence will be faved to government, and the military habits of the foldiers will be preferved; whilft they will have the benefit of better attention, both from the furgeon and their comrades. The furgeons themfelves, will have an opportunity of acquiring knowledge and experience; and will be employed in the proper line of their duty.

To correct abufes in the French departments, will occupy the attention of the infpectors in St. Domingo; with the greateft advantage to the country. It is aftonifhing with what eagernefs and acutenefs they purfued depredation, and what a variety of ingenious pretexts they formed, to attain their purpofe. Nor is it eafy to exculpate their leaders from fome fufpicion of countenancing thefe frauds.

The

The difficulties, with which any improvement was introduced, which propoſed reform, or the diminution of expence; lead to a belief, that they have an intereſt in ſupporting the impoſition. Dr. JACKSON, found unexpected obſtacles thrown in his way, by commanding officers. And it was after much trouble and difficulty that his plans were partially executed. Yet his ſcheme made a ſaving to this country of £.50,000 a year; no ſmall retrenchment in one department.

In this charge, I do not, without diſcrimination, involve the French officers of rank; this would be illiberal and unjuſt; there are amongſt them, men for whom I entertain the higheſt reſpect, and who deſerve well of this country; but there are alſo men, to whom theſe charges will fully apply, and who have enriched themſelves by the ſpoil of this iſland.—

HOSPITAL CORPS.

ANOTHER argument against general, and in favour of regimental hospitals, arises from the mode of attendance on the sick. It was impossible at Port-au-Prince, to procure female nurses for the general hospital. It became therefore necessary, to employ soldiers, who having escaped the endemic, or recovered, had more confidence than others. But although soldiers readily attend their own comrades, in the same regiment, they do not so willingly wait upon others. Besides, the regiments to which such soldiers belong, do not easily agree to let them continue in the general hospital. These difficulties, which were represented, gave rise to the levy of an Hospital Corps; from which all such men were to be taken. All the orderly men or nurses, were to be supplied from this regiment; and the military commandant was to see, that they behaved in a decent, regular, and proper manner. He was to attend to the complaints made to him by the physicians and surgeons, and to direct his men, in the most useful manner for the benefit of the sick. To perform these duties, it is evident

evident that the men who compofed this corps, fhould be regular, fober, and humane. No duties are more facred, than attentions to the fick; no duties require more ftrictnefs of manner, or greater decency and firmnefs of deportment. Such a corps, ought to be compofed of the beft men of the army, to whom this occupation might be given as a releafe from feverer duty, and a reward for good conduct; as the pay is better than in the ranks, and the fervice to be performed, not quite fo laborious as in the field. But inftead of forming this body on thefe principles, they were made up of outcafts from all regiments; either of men, whom difeafe had incapacitated for any duty; or of thofe who had abandoned themfelves to drinking and debauchery. Thus conftituted, they were much fitter for Botany Bay, or the hofpital of invalids, than any employment which required humanity or action. But to thefe men we were obliged to confign the nocturnal charge of the fick; on thefe, they depended for drink, and every other affiftance during the night. Many neglects muft have happened from fuch attendants. Had this corps been properly made up of well behaved and fteady men, it might be of great
service;

service; but in the manner it was constituted, could not be of any use.

Another great difficulty, occurred to the medical gentlemen who had charge of wards. There were not a sufficient number of assistants, to make up the prescriptions, or pay the necessary attention in seeing them taken. On many occasions, a ward containing eighty patients, had only the attendance of one medical gentleman, who was obliged to be apothecary, attendant, surgeon, and physician. It is true, we trained up some careful men to assist in making up preparations, but they could not be wholly trusted to their care.

Such were the labours that medical gentlemen had to undergo in this destructive climate. Others had some rest from their labours, they enjoyed no interval of ease; no cessation from their toils. The government at home were not to blame for this scarcity of assistants. They had sent out numbers, on the staff, and attached to regiments; but the climate had swept them away. One man who is once seasoned in that country, and can be depended on, is equal to a host of strangers, who themselves require assistance, and fall
victims,

victims, when their exertions are moſt required. Seaſoned men, ought to have every encouragement, and ſhould be ſought with diligence; on them only can the ſick, rely for aſſiſtance. Ere they are ſeaſoned to the climate they muſt acquire experience; and theſe two qualities are invaluable. If a ſervice does not reward merit, it will never have men of talents; they will deſert it; and employ themſelves, where they are rewarded. How many regiments and ſhips of war, have I ſeen in St. Domingo, without a ſurgeon or mate, from the plan of employing young and inexperienced men.—They fell victims themſelves, when their aid was moſt requiſite. But ſuppoſing, they had lived, what aſſiſtance could they have given to the ſick; a new ſcene was preſented before them, and unheard of deſtruction. The rapid progreſs of an aſtoniſhing fever, amazed their faculties, European practice was tame and feeble, former experience only tended to confound them; they were idle ſpectators, till they themſelves periſhed. It is true, that we muſt all begin to acquire experience, in a ſtate of comparative ignorance; but thoſe who command armies, ought to chuſe their medical aſſiſtance from warm climates; which, though they may not preſent ſcenes like St. Domingo,

Domingo, will at least prepare the physician, more than the schools of Europe. I am sure I shall not be contradicted, when I assert, that it is absurd to send out physicians from London, to combat the diseases of St. Domingo.— The requisite knowledge for this purpose can only be acquired on the spot, after a long, painful, and accurate attention. Out of seven physicians, all highly qualified in their profession, who were destined for the St. Domingo establishment; only two were doing duty when I left that island. Dr. CLEGHORN, fell a victim to the fever at the Mole. Dr. FELLOWS, after struggling long, to perform his duty, was obliged to retire, in an almost hopeless state. Dr. CAVE, has been obliged to return also; Dr. MASTER, who in a state of illness, nobly continued his labours, underwent much sickness; he and Dr. HENDERSON only remain of the seven physicians who were meant for the island.

Instead of sending medical men from England to those islands, where they have little chance of exercising their profession, or even of preserving their lives; physicians of character ought to be encouraged, from Jamaica, Barbadoes, and other hot climates, where similar

similar diseases reign, and where they have been already seasoned. Those who have followed the army and acquired experience, are inestimable, and should receive every encouragement. To a knowledge of the climate, and its diseases, they add another valuable acquisition, the knowledge of the habits and manners of soldiers; the most necessary knowledge to a military physician or surgeon.

Let the most celebrated physicians of Edinburgh or London, be sent out with an army to a warm climate, like St. Domingo; without being previously seasoned, and acquainted with military habits; and I believe they would feel, and acknowledge, the truth of these remarks. It is astonishing that any other mode of recruiting this establishment, should ever be thought of. The army itself, generally supplies a great number of men of good abilities, who have braved the climate, and seen a great deal of practice. To these, in proportion to their abilities, a just preference should be given; when they are not to be found, it is then fair and proper to look out for others, who may be competent to fill the station. But in the West Indies, or in such islands of them, as have a large army,
there

there ought to be a *Medical Board*; to examine such candidates as may offer themselves on the spot, and may be requisite to nominate. It is there only, in actual practice, and doing actual duty; that a judgment can be formed of the abilities or industry of any one. Such a board might be composed of two inspectors, two physicians, and two surgeons; who would regulate and fill up vacancies, and recommend the proper people for advancement. — The recommendation of such a board, to the commander in chief at home; ought to insure the candidate, whatever they recommended. In this manner, the service would never be in want of proper assistance; and the diligent, industrious and able candidate, would be rewarded. Men will not chuse to enter a service, and forsake other pursuits, on vague and uncertain grounds; nor is it fair to expect it, without some assurance of benefit. But on the present footing, nothing can be promised, till an answer is obtained from home. In the mean time, the service suffers, by the want of assistance, which might be procured; if the inspectors on the spot, and who alone can be judges, enjoyed sufficient power. I could adduce many proofs of inconvenience and loss to the service, from this management.

R   Boards

Boards are a council, for the commander in chief, to advife him on points, in which he cannot be fuppofed himfelf perfectly converfant. Of this kind are medical appointments; of which medical men, ought undoubtedly to have fome direction. They are to take care, that no improper perfon, fhall fill any medical fituation in the army; they are anfwerable to the commander in chief for their recommendations, and are in fact to advife him on the medical department. Thus far their powers extend, and thus far they are proper. If it be right to grant thefe powers to a board at home, it would be ftill more proper, to grant them to one abroad. The board of England is competent to regulate all the bufinefs of the three kingdoms, becaufe the difeafes of thefe countries are nearly fimilar; and the courfe of education purfued in our medical fchools, qualifies their pupils for this fervice. Abroad, it is otherwife. No courfe of lectures, no reading, can qualify for that fervice. A long experience, and accurate obfervation, can alone entitle the practitioner to any confidence. The difeafes are too rapid for delay, too dangerous for doubts. The phyfician muft at once decide, or the patient is loft for ever.

A very

A very proper caution, and a very proper respect to an useful form, have been too rigidly adhered to, in the conduct of Medical promotions. In the origin of degrees, or medical graduation, the chief point of importance, was, that by holding out a certain number of profeffional examinations, through which the candidate must pass, he would be obliged to possess a liberal education, and pay strict attention to his pursuits. When he had passed these trials, he came forth sanctioned to the public, by the approval of a learned body, supposed, impartial judges of his merit. The public became thus guarded against empiric pretenders, against improper and illiterate practitioners. No liberal man will deny these useful and necessary forms, his warmest approbation. It must be confessed, that many universities, departing from the dignity of this form, have prostituted their sanction, to any person, who could purchase their seal and signature. It is therefore necessary, that some colleges, maintaining the purity and intention of the original form, should gain to their candidates a preference.

The benefits of this general protection, against empirics, were very properly extended to the army of late years only. But the rule has been adhered to, with more than ordinary rigidity. The line of phyſician to the army has been confined to the graduates or bachelors of the Engliſh univerſities only; ſome exceptions have been made in favour of Dublin. I have ſaid, that the form itſelf is highly proper, but I ſhall go farther, and ſay, that on a fair liberal conſtruction it cannot be too ſtrictly adhered to. But making every allowance for the celebrity of other univerſities; I believe it will hardly be queſtioned, that EDINBURGH, is at this moment the FIRST MEDICAL SCHOOL IN EUROPE. If this is true, as I believe will be acknowledged, upon what principle are the graduates of that college excluded from being phyſicians to the army? unleſs they are alſo licentiates of the London college. It cannot be fairly ſuppoſed, that Edinburgh, anxious for the fame of its ſchool, is leſs ſtrict than the London college in its examination of candidates. If the other univerſities are leſs ſcrupulous, it is no good reaſon for claſſing Edinburgh with them, under a general prohibition. An examination at London, will hardly inſpire any candidate with ſudden knowledge. And I imagine, it will

not

not be denied, that a ftudent from the Edinburgh School, is as likely to underftand the theory of Medicine, as a pupil of Oxford, Cambridge, or Dublin. Fortunately, fcience is not confined to any one univerfity, but may be acquired in all, by genius or induftry.

But if a reftriction is neceffary, and I am of opinion it is; let the great feminaries be privileged to offer candidates, viz. The Two ENGLISH UNIVERSITIES, DUBLIN, and EDINBURGH, let their graduates, without any further examination, be eligible, for the office of phyficians to the Forces. But it is not furely neceffary to oblige thefe graduates to pafs at London. The urgency of fervice, and the neceffities of war, fometimes require that the moft pofitive rules, which guide, in time of peace, be laid afide for the benefit of the fervice. What will do in time of peace, will not anfwer in time of war.

Thefe remarks apply to every department of the army. Promotions muft be made for the purpofes of war, which in time of peace, would very juftly be reprobated as improper; as an infringement on the general rules of the army.

army. But thefe cafes of neceffity ceafe, when the urgency which gave rife to them no longer exifts. The EAST INDIA Company, the moft extenfive and opulent commercial body in Europe, have laid down very juft and proper laws to regulate their fervice in time of peace; but in time of war thefe laws give way to cafual urgency and unforefeen neceffity. If it becomes neceffary then on fervice, and in time of war, to break through eftablifhed rules, in the great body of the army itfelf; is it not to be expected that in every leffer department of it, urgencies and neceffities may arife, which juftify an infringement on general rules, that have not provided for thefe cafualties.

Let us fuppofe, that an army, well appointed in the medical department, fets out for St. Domingo; but foon after its arrival there, is deprived of its phyficians and furgeons, who perifh in the fever; let us further fuppofe, that this army continues fickly, and every day, more and more requires affiftance. Jamaica is in the neighbourhood, with many experienced practitioners; but alas! they have not undergone an effential form, they have graduated at Edinburgh or Dublin, and therefore

fore are not eligible. Would this reasoning guide the commander in chief; would he permit his men to perish, because no licentiate of London could be called to their assistance?—

But independent of this difficulty, another serious objection would oppose his procuring assistance. A practitioner in Jamaica or St. Domingo, unconnected with the army, and engaged in other pursuits, would not readily be induced to forsake these, and engage in a service, for a temporary benefit; some inducement must be held out, some permanency offered; but this neither the commander in chief or the director of hospitals can positively do. In these circumstances, which are not unlikely to happen, and which I have seen nearly verified; the army may perish without assistance. From this statement, it would appear, that in certain circumstances, it is just, expedient and proper, to break through rules, which have made no provision for urgency and necessity; for the great casualties and calamities of war. The rule may be proper in time of peace, but does not apply to war.

But independent of the diplomas of univerfities, which certainly prove, that a man has walked through the formalities of his education, as barrifters eat their commons; public teftimonies of another kind might be fometimes admitted in favour of a phyfician to the army. Of this kind, are the publications of medical men, their profeffional character and fuccefs. I fhould have conceived, that Dr. Jackson's book on the Fevers of the Weft Indies, would have entitled him to be a phyfician to the army; if he had no other teftimony, and had never feen London, Edinburgh, or Dublin.

I think upon the whole, that the rule of making licenciates only, phyficians to the army, is too rigidly adhered to, and that the power of the director of hofpitals is too limited; and that both may be productive of bad confequences to the army. I am not fure, whether it would not be for the benefit of the army, to make it a ftep, to the furgeons, after a long and approved fervice, and thus make it a military graduation. This would hold out fomething to men of experience, to continue in the fervice, fomething to reward their toils and labour; at prefent, it muft be confeffed, that
there

there is no encouragement, to bind a man of talents to the army. I believe, that this was once a plan in the service; and if strictly but fairly regulated, would prove very beneficial.

Having now gone through this miscellaneous section, I shall proceed to say a few words, on Wines, and some further remarks on Diet, in which some observations will be offered on smoaking; and bilious complaints.

SECT. IV.

Confiderations on Diet, adapted to the Climate—Claret and Madeira compared—Diet recommended by the Author—Smoaking confidered—Its Effects—Flannel examined—Cotton Shirts recommended——Bilious Complaints—Caufes of Bile—Cure—Of the Prickly Heat—Its Caufes and Nature—It is not dangerous or critical—Bathing not dangerous when it is out.—

IN the general remarks offered on Diet, I did not enter fo particularly into the fubject as might be requifite. I fhall here obferve, that I think fome changes might be introduced with real advantage in the diet of our foldiers in the Weft; particularly in their drinks. Rum, in the manner they get it, muft poffefs pernicious qualities; it is of the very worft kind, ftrong and ardent. Moderation is not the moft confpicuous virtue among foldiers. I think good porter might be fubftituted with great advantage inftead of rum:—It ought to be of the beft quality and bottled in that country:—Of this it would not perhaps be too much,

much, to grant each foldier three pints per day. Porter is more nourifhing, and lefs pernicious than rum or brandy; it has not that active ftimulus, which by repetition foon exhaufts the excitability, or gives vigour to morbid caufes. This might be occafionally varied with fpruce beer, or a pint of found Madeira. On fome occafions, fuch as, when a centinel is expofed to a cold, or rainy night, a wine glafs full of good brandy may, I think be given with advantage; one as he enters on duty; and another when relieved, before he lays down. I do not imagine, that this would be injurious, in thefe particular circumftances. It would tend to fupport the vafcular action, which the application of cold is apt to diminifh; and this diminution of action feems favorable to the invafion of difeafe. The excitement of a glafs of brandy would not be fo great as to bring on any formidable collapfe. The porter fhould be fo divided, that no more than a pint be taken at a time; in larger quantities, it brings on drowfinefs, and favours the production of bile. One pint might be diftributed in the forenoon, mixed with water; another at dinner-time, and a third at bed-time. This would be fufficient for the fupply of moifture moisture and nourishment, and would in no degree, produce intoxication.

Thus much, regarding the soldiers; I shall now consider the diet and drink of officers, who live in a manner somewhat different. And first as to wines :—The chief of these used in the West Indies, are Madeira and Claret; port being reserved only for the sick. Good, sound, old madeira, is no doubt an excellent wine in that climate; in this condition the volatile, ardent and spirituous part has in a great measure fled, and the body, consisting of a great basis of water and the juice of the grape remains. But few officers drink this quality of madeira, because it cannot be procured; they are obliged much oftener to use a secondary kind, into which a good deal of brandy has been thrown. This sort of madeira, is nothing more than a mixture of brandy; and the drinking of it must be very pernicious, and afford many opportunities, for the attacks of fever. Claret is less subject to adulteration, and when found, and of a good body, appears to me the fittest wine for the East or West Indies. It possesses more of the nutrient grape, and less of the

spirituous

spirituous part than madeira; and it is not
by any means so apt to intoxicate; it possesses
besides an useful laxative quality. Supposing
then madeira and claret, to be of equal good
qualities in their kind, I should give the preference to claret. I believe it less injurious
to the constitution; less stimulant and heating; and more temperate and nourishing.
Fashion and accident frequently guide mankind in the choice of the most important
things; no wonder that they should rule their
luxuries, which are their offspring. In the
East Indies, universally, a preference is given
to claret; in the West, madeira bears the palm.
The claret used in India, under the name of
English claret; or claret for the London mar-
ket, has always appeared to me the best kind
of it; it has a full body, and is somewhat
more powerful, than what the French themselves used. I believe, it will be generally
allowed, that claret neither intoxicates nor
heats the body so soon as madeira. These
two circumstances, are in my opinion a good
ground for a preference to claret: but many
are deterred from drinking claret, from an idea
that it has not stimulus enough for a relaxed
stomach; or that it is, what they call too
cold. I believe this objection to be partly
fanciful;

fanciful; but to prevent this danger, a few glasses of madeira may precede the claret, so as to heat and warm the stomach; or what would answer equally well, a half wine glassful of brandy; with this foundation, the claret may be taken without any danger to the stomach. There are many who believe that pretty hard drinking contributes to their safety in a warm climate. I am by no means of this opinion, for I think in its consequences, it subjects them more readily to the invasion of fever; but moderate living with a due proportion of solid, and liquid nourishment, contributes to the maintenance of health, and preserves the constitution from disease. It is very difficult to draw the line accurately; but I should think it a good general rule, to eat small portions of animal food; to eat soup or broth, and to let vegetables have a place in our repast. As to drinking, we may perhaps do well not to exceed what promotes cheerfulness and a social spirit, without intoxication. The quantity which in different constitutions produces this disposition, is as various as the constitutions themselves. But I should imagine, that a bottle of claret after dinner, when a person sits two hours, cannot prove injurious to the constitution; for in
this

this climate there is a constant perspiration going forward, which exhausts the body, and diminishes the fluids, unless a constant supply is supported: and as the proportion of animal food ought to be diminished, from which the chief nourishment is extracted, this mode of deriving it from fluids, becomes more necessary and useful. Perhaps a life of absolute sobriety, would not conduce to health in the West Indies; the system would become too weak and languid; and obstructions might happen; so that sobriety itself might be an error; but to this error few officers are likely to fall a sacrifice.

But it must not be supposed, that I am encouraging debauchery, or drunkenness, when I recommend a rather generous manner of life. There is a great difference, between living well, and living freely, between moderation and excess; but we observe, in warm climates, that the natives, who live entirely on vegetables, and to whom religion forbids, wine and animal food, are weak, timid, and incapable of exertion. They do not attain the strength or activity of their neighbours, who indulge in these articles of food. The Gentoos, or Hindoos, are not so athletic as the

the Mahometans or Persians, who are not restricted by the same laws of abstinence. Hence, it is fair to conclude, that moderate portions of wine conduce to vigour and the maintenance of health.

It is a custom in the West Indies, founded on sensation, to drink during the forenoon, and the day, some diluent, refreshing drinks; this is done in obedience to thirst, a very imperative sense. Weak sangoree, or a drink made up of sound madeira, water, acid, and sugar; lemonade, tamarind water, and such compositions, are cooling and pleasant; and may contribute to health by supporting perspiration. Perspiration greatly conduces to health: it preserves, by the evaporating process, a great coolness in the body; it relieves the vessels from the distention of the fluids, and permits the expansion by heat to go on without pain or detriment. It diminishes the saline and stimulant part of the blood; and it may throw out of the body the miasmata themselves, which excite and cause fever. The obstruction of this most useful discharge, must be productive of the worst consequences; but it can only be supported, by keeping up a regular supply of fluids, which these mild diluents

diluents very amply do. It is sometimes astonishing, how quickly it is produced after a drink of this kind; it bursts forth almost immediately. Whenever I found the perspiration diminish, and my skin becoming parched or dry, I used to re-establish it immediately, by a draught of sangoree, or lemonade. A free perspiration, is the surest preservative of health in a hot climate. In carrying on duty at St. Domingo, I was exposed to a good deal of riding, being generally six or eight hours on horseback every day; and consequently I perspired very freely.—I never, wore flannel, but made use of cotton shirts. It was not unusual with me to shift five times a day; sometimes oftener, each shirt being drenched in perspiration. To this profusion of it, I attribute my safety, amidst so much exhalation of miasmata, for so long a time. I drank freely of lemonade, sangoree, tamarind water, weak wine and water, and other diluent compositions; and when night came, I was always prepared to enjoy my repose.

The tea breakfast, used by us in the West Indies, appears to me less suited to the climate, than the breakfast of the French; though I think, they rather incline to excess. What is called a second breakfast in our islands, does not seem to me calculated to support health. But I do not object so much to tea itself, on the ground of its being, what is called nervous, as to the quantity of warm water we drink with it; which is certainly relaxing, and which tea is not calculated to correct. Now it is of the utmost consequence, to preserve the powers of the stomach, as entire as possible; for the vigour of the body increases, or diminishes in proportion to the powers of this most essential organ; from which as from a center, strength and nourishment are propagated to the rest of the system.— But warm water is not calculated to strengthen the stomach, or add to its energy; on the contrary, its long continued use must relax, weaken, and impair its powers; nor is the rest of the breakfast calculated to inspire vigour. The butter is seldom good, nor are oils easily digested. The breakfast which I used, and found light and nourishing, was made up in the following manner. With bread, I used fresh eggs, or a small portion of tender beef steaks, or broiled pigeons, or a slice of beef; and instead of tea and hot water, I drank cold water with a fourth part claret. This breakfast I always found

light,

light, and eafy of digeftion. The firft part of it, afforded fufficient nourifhment, without creating bulk; and the cold water and claret braced the fibres of the ftomach, and gave them ftrength. I placed great reliance, on the water's being very cool; to effect which, I had it in a jar of a loofe clay texture, through which the water perfpired, in the day time; and in the night it was expofed to a ftream of air; fo that in the morning, it was cool and pleafant. After this breakfaft, I found myfelf light and vigorous, and equal to the duties of the day. I muft confefs, I think it more falutary, than tea, hot water, and rancid butter. I never indulged in fuppers, which I think are too heavy meals, for a hot climate. The ftomach muft be affifted by exercife to perform its functions; it is not fo powerful as in Europe; and as it has not the advantages of exercife to affift it after fupper, that meal may be laid afide with advantage.

Some caution is neceffary in the ufe of fruits, in which ftrangers are apt to indulge immoderately at firft. I think the ufe of them is more falutary at breakfaft, than after dinner. They are applied more clofely to the coats of the ftomach; there is more exercife to digeft them;

them; and the taste itself is more pure, and is a better guide than after dinner. Oranges, melons, pine apples, plantains; in moderate quantities, are good and safe fruits; there is another which is not inaptly called vegetable marrow, from its resemblance to that animal substance, which may be used with great safety. —I have seen a preparation of it, with lime juice and sugar, which resembles in taste strawberries and cream. It will not be a bad rule to restrain the appetite for some time, in the use of these fruits, gradually allowing more indulgence, till they can be used with perfect safety. That the stomach may not give way too much to a new stimulus, it will be a proper rule to take a little brandy after using fruits; and in this manner, I think they may be taken with safety and advantage.

Nature, in every climate, seems to have produced, what is more immediately useful and grateful to its inhabitants.—In most countries we not only find the necessaries of life; but that kind of them most suitable to the nature of the climate. Nothing can be so grateful to the thirsty palate, as the mild acids of the orange and tamarind; nothing more luxuriant, than the delicate flavour of the pine apple or

† the

the mangoe. These happily abound, where they are most requisite and grateful. Every warm climate abounds in them; the poor and the rich may banquet at nature's luxuriant table.

I have seen a practice pursued in the army, which I think very improper. I allude to the giving the men ardent spirits previous to their undertaking fatigue. Nothing can be more preposterous than this practice; instead of enabling men to bear fatigue, it wholly unfits them for it. If it is meant to create a temporary frenzy, in a desperate attack, it may perhaps answer the purpose; but unless the enterprize is effected in a moment, we will be disappointed; the languor soon follows the excitement, and renders them passive; and it has been already shewn that great walkers, and men who undergo the greatest fatigue, are those who live moderately, and during the performance of those exertions use water chiefly.

Great portions of animal food exhaust the system, like the use of ardent spirits; the excitability is wasted, and an inclination to sleep induced. Previous to the undertaking of any serious service, where fatigue is expected, the canteen of every soldier on parade, should be examined, and filled with lemonade or water. No spirits or mixture of spirits should be permitted, till the service is performed, when greater indulgence may be granted. In the state produced by spirits, soldiers either become rash and disobedient, or obstinate and cowardly. In a state of sobriety the influence of habit and discipline, will make them follow their officers and obey orders.

I come now to speak of another luxury connected with health, which officers and soldiers use in the West Indies, viz. SMOKING TOBACCO. This is an artificial luxury, few men naturally take to smoking tobacco; because it is at first highly unpleasant, and a taste for it can only result from perseverance and habit. Smoking in India is a real luxury; the hookah contains, the most grateful odours of the Eastern spices; the tobacco itself is of a particular delicate kind, and is so involved among other ingredients, that its peculiar smell or flavour are not perceptible. I have seen European ladies, sit in the rooms, where gentlemen smoked, to enjoy without trouble, the fragrant smell, the hookah diffused;

fused; nor do I recollect, that I ever saw it unpleasant to any stranger. The smoking of tobacco has no claim to this character; it is almost universally disagreeable on first trial; and it requires no small effort to persevere, and render it pleasant.

It is not easy to ascertain, what first led to the origin and commencement of this practice. Perhaps an opinion of its medical virtues, induced some, whilst others commenced from mere idleness. The languor of heat, and the inactivity of cold, have equally produced smokers, because the effect is nearly equal. In Asia, the more elegant luxuriant methods are employed, which amuse and excite, the gay and volatile inhabitants of these fertile regions. The Turks delight, in the pleasant fancies, and delirium of opium; whilst the more phlegmatic, and less irritable inhabitants of Kamschatka, and Holland; use the more powerful stimulus of tobacco, which alone can rouse their torpid and lethargic habits. In the West Indies, the only improvement on the Dutch method, is the use of the segar—which is a milder kind of tobacco than the Virginian. In some places it undergoes particular management; it is exposed to the sun,

after

after being repeatedly washed in water; so that the strong essential oil is dissipated and evaporated, and the leaves rendered milder. After this preparation, they are rolled up into tubes for use. The Spaniards are famous for their segars, which by connoisseurs are supposed to excel all others.

There cannot be a doubt, that the use of smoking has been pursued for years, by men who have not materially suffered from it; just as men pursue a course of drinking, without seeming to receive injury. But this is a negative and very doubtful proof of its healthfulness. When we attend to the effects of smoking, we find, that after proving a very confiderable stimulus, great languor is induced. For tobacco produces nearly the same effects as ardent spirits or opium. It exhausts excitability, and exposes the constitution to the action of morbid powers. The worst effect of smoking tobacco, is the loss of saliva or spittle, which it produces. This not only weakens the system, by creating an unnecessary evacuation, but proves also highly destructive to the digestive powers of the stomach. For there is no doubt, that the saliva, which forms a large portion of the juices

juices which blend and subdue the ingesta of the stomach; possesses great power, in the process of digestion. It has been proved in Dr. BLACK's experiments, and in those of IRWIN, that the muscular powers of the stomach have less influence on digestion, than was generally imagined. The liquors which seem to act as menstrua to the food, appear to carry on the digestive process, with little reliance on the muscular structure of the organ, where it is carried on. The expulsion and propelling of the mixed mass, seems wholly to depend on the muscular fibres of the stomach; and something must be granted to their action in the digestive process itself. For digestion is seldom well performed, where we have reason to suspect relaxation in the fibres of the stomach; and we find, that we improve the appetite and digestion, when we succeed in removing the relaxed state. But whether this lax state of fibre, affects digestion from the mere effect on the muscular fibre, or by changing the nature of the gastric juices, so as to weaken their powers; cannot be easily ascertained. At any rate, if it be admitted, that the saliva, is necessary or useful in digestion, the waste of it by smoking, must prove highly injurious to that process. And in fact,

fact, I have seen it so repeatedly; few of the eminent smokers, are able next morning to enjoy their breakfast, and make no great figure at dinner. I had an opportunity in my own house at St. Domingo, of remarking particularly the effects of smoking. A very amiable young man, lived with me for some weeks; who was uncommonly attached to his segar. He began as soon as he was out of bed, and continued it all day, with little intermission, until bed time. The consequence was, he lost his appetite almost entirely, he could not digest any solid food, he lived on soup, and other light articles; he was thin naturally, but became more emaciated, from the continuance of smoking. This marasmus or wasting, and the loss of appetite; I attribute entirely to the great expenditure of saliva by smoking.

I have seen in many people of a costive disposition, that the stimulus of smoking, communicated from the mouth to the stomach and intestinal canal; has produced action in them, and procured a stool; many people smoked for this purpose only. The soldiers smoked for pastime, and to obviate the effects of rain. It raises the pulse, and
maintains

maintains a confiderable excitement in the fyftem; in this way it acts in the fame manner with ardent fpirits or brandy; and may in moderation be really ufeful. Like other ftimuli, the quantity and duration muft be increafed; otherwife fmoking lofes its effect; at length it becomes a neceffary habit, and enflaves the conftitution.

Upon the whole then, it feems to have had its origin, from laffitude and idlenefs, and to have been perpetuated in climates, where thefe difpofitions are created from extreme heat or cold. It would feem on the whole to be inimical to health, by wafting the faliva, fo neceffary to digeftion, and thus weakening and diminifhing the appetite. But that in particular fituations, the ftimulus of it may be moderately ufed with advantage, to promote the expulfion of the fæces, and to refift the influence of cold, by fupporting vafcular excitement. But in general, it is ufed too freely in the Weft Indies, to an extent, and duration, that brings on debility; and leaves the fyftem expofed to the attack of difeafe. It is unfortunately introduced in the hours of drinking, when the tafte is vitiated, and our fenfations lefs perfect, and it is not uncommon

common in these circumstances to see it continued for five or six hours. The consequences may be easily imagined; languor and debility are experienced throughout next day; the appetite is destroyed and all vital energy diminished. No situation can be imagined; more inviting to the invasion of Fever. And I have seen many instances of an attack the ensuing day, after excessive smoking; to which no doubt excessive drinking had been conjoined.

Before I close this section, I shall add a few observations on Flannel, as used now very generally by our troops in the West Indies. The great use of flannel next the skin, has appeared to me, to be the maintenance of an artificial, but uniform climate, which prevented slight vicissitudes of weather from affecting the soldier; when necessarily exposed to it. Another use, which may be almost reduced to the former, is the absorption of perspiration, which in a linen shirt, would be applied cold to the body, and check the energy of the vascular system. That flannel, may in some degree produce these effects I believe, but I imagine it has contributed much

much lefs to the health of foldiers in the Weft Indies, than may be generally imagined. It appears to me, to debilitate, by exciting too much perfpiration, and keeping it inceffantly up; and to render the foldier lefs fit for the cafualties of fervice, by creating a great artificial fenfibility, alive to any confiderable change. Befides, the foldier feels incumbered, hot, and uneafy in his exertions. But a worfe confequence ftill, is apt to refult from it; it may become the nurfery of difeafe, by retaining the exhalations from the body, and having them again applied with the chance of abforption It is well known, that no attention can prevent the foldiers from wearing it too long; when it becomes a foul nurfery of uncleannefs. They cannot carry with them a fufficient number of thefe fhirts to change as often as they ought; in fact, to be clean or comfortable, they ought in a warm climate to fhift every day.

Inftead of the flannel jackets or fhirts, which are heavy, hot, and uneafy; I would recommend, what I think would anfwer the purpofe much better, without any of the difadvantages mentioned; I mean cotton half fhirts shirts without sleeves, in the manner worn in India under the denomination of banian shirts. These would be easier carried about, the soldier could pack up a good number of them, as they do not fill up much room, by which means he could shift oftener, and keep himself more clean. These would absorb the perspiration, and create less heat; they would preserve the soldier from the danger of vicissitude, and dissipate his vigour less, than the flannel. He might always have a sufficient number for the necessary change. During all my labours at St. Domingo, I never wore flannel; cotton shirts were my only protection, and I found them sufficiently comfortable amidst every vicissitude.

I have now discussed almost every thing I judged of any importance to touch, relative to the health and preservation of troops in the West Indies; a subject of great importance, were my abilities equal to the weight of the discussion. I have fairly stated what I saw, and the conclusions I drew on the spot after much experience. I think it is the duty of every practitioner to come thus forward, and contribute to our stock of facts, and the mass of opinions. Subjects appear in different lights to different physicians,

physicians, and in the variety of positions, we may at length gain the most distinct view of which they are capable. Before I come to the Appendix, which is only in proof, that the reasoning I have employed has obtained tacitly in medicine, I shall subjoin a few remarks on BILIOUS COMPLAINTS; and what is called the PRICKLY HEAT; both very troublesome diseases.

OF BILIOUS COMPLAINTS.

NO subject on which, professional men, or patients, speak more inaccurately or loosely, than on what relates to the secretion, or production of Bile. Bile in warm climates, has the same general agency ascribed to it, that in colder regions belongs alone to the nervous system. Every complaint which is any how complex or inexplicable, is *bilious*. Diet is divided into what is bilious, and *not bilious*; as if it was instantly converted in the stomach itself, into this fluid. It may be proper then to take a short view of this subject, so as to speak on it with more accuracy. It is hardly necessary to premise, that the bile

is

is not a liquor produced or generated in the stomach itself; at any time, or in any circumstances; that diet of any kind, can only produce a larger quantity of this fluid, by acting on the liver and its veffels; through its confent or vicinity to the ftomach. That its excefs only can be a difeafe, as a certain quantity of it, is effential to the purpofes of digeftion. The liver is the feat of this fecretion alone. It does not exift at all in the ftomach, but is wholly derived from this gland. But from the neighbourhood of the ftomach to this organ, whatever affects it will in fome degree affect the liver alfo.

Bilious affections, or an increafed flow of bile, occur either fingly and diftinct, or combined with other difeafes; I fhall confine myfelf chiefly to thofe accumulations of this fecretion, which arife more especially in confequence of heat. The circulation of the blood in the liver, is carried on in a particular manner, a vein is made to perform a double office; and the movement of the blood is more flow, than perhaps in any other organ of the body. The general effect of heat, muft be felt here, as in every other part of the fyftem, and as the circulation is ufually quickened,

ened, larger quantities of blood will pass through the hepatic vessels in a given time. Heat besides, may stimulate the glandular structure of the liver itself; it certainly imparts sensibility and irritability to every organ. There is reason to believe that most of the secretions, are increased by heat. The semen is more copious in warm climates, than in cold ones, this, and the irritability of the whole system form one cause of the "Cupido Veneris," so remarkable in tropical regions. We are led to conclude, that heat itself, independent of other causes, operates powerfully in causing large secretions of bile. Because such secretions, are seldom observed in cold climates, and because, when they do occur they seem to arise from causes, which operate in a similar manner to heat, or rather produce heat itself. Thus the immoderate use of ardent spirits, debauches of wine, and violent exercise, are known to produce great secretions of bile. Whatever remains for a long time in the stomach, and resists the process of digestion, is apt to bring on a large secretion of this fluid. Of this kind are veal, lamb, eggs, salt meats, pickles, cheese, oils, and fat meats. Large quantities of porter, likewise increase the secretion of bile.

bile. Smoaking tobacco, or whatever ſtimulates the general ſyſtem, or more eſpecially the ſtomach; contributes to this ſecretion. Theſe, by making the fibres of the ſtomach act, either with unuſual vigour, or for an unuſual time; by ſtretching them, or by creating bulk, and preſſing on the liver; in all theſe ways, perhaps, contribute to an increaſed ſecretion.

Unuſual ſecretions of the biliary fluid, are emphatically called, "The Bile," and conſtitute often, a very troubleſome diſeaſe. Many ideal doctrines, and abſurd notions, are entertained on this ſubject; people imagine, that what they eat or ſwallow is converted into this fluid in the ſtomach. This opinion can only ariſe from want of anatomical knowledge. It has been ſhewn, that heat of itſelf is apt, and in fact does produce, a tendency to large ſecretions of bile, in warm climates. If heat produces this effect, by ſtimulating the vaſcular ſyſtem, and expanding it, how much more powerfully muſt it act, when aided by other cauſes, which increaſe irritability and ſenſibility; and which add to the circulating maſs, that muſt paſs through the liver; or ſolicit in that organ itſelf, a more vigorous action,

tion, by the sympathy and contiguity of the stomach. Heat assisted by these causes, produces enormous, and morbid secretions of bile. To the co-operating causes, may be added nausea, one of the most powerful agents in the increase of bile. The causes which increase the secretion of bile from the liver, may be reduced to the following

HEAT,

Nausea or vomiting, in whatever manner excited;
The immoderate use of ardent spirits or wine,
Violent exercise, especially after meals,
Viscid food, whether solid, or gelatinous;
Heavy meals, over distending the stomach.
Fat or greasy food, oils, and acids.
Depressing affections of the mind.
A morbid sensibility in the liver itself;
Obstruction in its vessels.

To one or other of these causes, may be referred all the bilious cases, I have ever seen. In some, several of these causes combine, and render the disease very obstinate.

Having now enumerated the causes, which produce large secretions, let us attend to the symptoms, which they commonly bring on in the system.—The first is, usually, a general sense of weariness, with a slight aching in the bones; and a desire to recline, with a constant inclination to sleep. The appetite is impaired, or becomes suddenly voracious; but a preference is given to hot dishes, with large quantities of spicery; such as currie.—When the patient has eat heartily, the symptoms for a time disappear; and lead him to believe that he is well. The skin becomes dry, and a peculiar uneasy heat, is felt in the palms of the hands and soles of the feet. The eyes are moved with uneasiness and pain, a general languor prevails over the body. Perspiration in general is much diminished, the face alone appears moist; now and then a sudden burst hot and disagreeable affects the hands. The tongue is covered with a deep yellow tinge, a head-ach comes on, the vessels of the admata, wear a slight yellowish colour, sometimes deep; the patient is restless; anxious and uneasy; sleep is turbulent, interrupted by irregular recollection and slight delirium; and the head-ach itself encompasses more particularly the eyes and sockets. The

nausea

nausea is especially distressing in the morning, on the first attempt to get up; the pulse becomes very frequent and the heat intense. The belly is most frequently bound, though now and then a diarrhœa occurs, with a particular scalding sensation at stool. These symptoms continue for three or four days, till the bile is carried off; they are more or less violent in different constitutions, according to the nature and circumstances of each; and perhaps, the absolute or positive quantity of bile in the stomach. Bile even in its most natural state, and quantity, proves stimulant, and excites the action of the intestines and stomach; it must prove still more so when applied in large quantities, and when the stomach and intestines are in a more irritable state. Perhaps too, that these secretions made from larger supplies of blood, and in a state of acute sensibility in the organs, are in their own nature more stimulant and active. We know, that certain circumstances of the organs, render the secretions much more vigorous and stimulant, giving them entire new qualities. The saliva of the dog, is perfectly harmless in the healthy state of that animal; but when the secreting organs have undergone a change, in the madness of this faithful attendant,

tendant, this very saliva becomes one of the most vigorous and formidable poisons. Again, the state of the organs, being changed, the effect of any fluid on them, supposing it unaltered, will be very different. In the Remittent, the irritable stomach rejects its natural and mild juices; the semen passes through the flaccid penis, without any sensation; but when it is full of blood, and in a state of temporary inflammation, the same fluid creates convulsive motions, and pleasing sensations. If the bile then should not be altered in its qualities, but increased in quantity, and applied to the stomach in a certain condition, it will produce morbid effects. But there is reason to be persuaded, that no secretion is ever increased or diminished, without being changed in its qualities, because the condition of the secreting organ is changed; and therefore it is not improbable that something unusual is produced in the nature of the bile, when it is secreted in large quantities. All our organs are apt to be thrown into action by any unusual stimulus, though apparently very mild. But if the condition of the stomach be more irritable, and the quantity or quality of the bile is altered, are we not to expect a very vigorous action of that organ; and in fact, we find

find it fo—unfortunately too, this very action of the ftomach to relieve itfelf, by its vicinity and confent with the liver, affects this organ, and induces larger fecretions. Natural vomiting however, affords a temporary relief; but as the bile is again flowing into the ftomach, the head-ach and naufea return; and the fame procefs is repeated feveral times before any permanent relief is obtained. We feldom attain repofe, until the bowels are opened, either by the bile itfelf, which is not uncommon, or by means of medicines. This progefs of bilious fecretion, is what is termed in the Eaft and Weft Indies, a " Fit of the bile."

I have now given the general hiftory of its fymptoms; they are not fo violent in moft cafes, as I have defcribed them, though I have feen all the tranfitions mentioned, occur in many cafes; and not unfrequently in myfelf. The languor, drowfinefs, heat in the hands, and lofs of appetite, generally warn the patient of the approaching difeafe; and if means were early ufed, it is probable, that no ferious accumulation would ever happen.

To prevent morbid fecretions of bile, the occafional caufes muft be avoided. All food or drink, which from their nature are apt to remain long in the stomach, should be very sparingly, or not at all used. The stomach should never be over distended either with solids or fluids—for this very distention unquestionably creates bile. Suppers, smoaking, and ardent spirits, by throwing the stomach into action, favour the production of this fluid; and should be carefully avoided. Very young gelatinous meat, resists in a remarkable manner the action of the stomach, and therefore favours the increase of bile. Pork, veal, butter, cheese, smoaked beef, salt meats, resist the powers of digestion in hot countries, and should be avoided by those, who are subject to biliary disorders.— The diet should be light and easy, composed of fully ripened fruits; and large portions of vegetables, with a small quantity of animal food; such as roast mutton or beef, or fowls. —White wines, especially sound madeira, from a long experience, I can venture to recommend in preference to the red; of these claret appears the least hurtful. Beer and porter must be sparingly used, as they seem in such cases to be very prejudicial. I have indeed seen cases, where porter created a diarrhœa; in these peculiar circumstances, it was an useful remedy. Riding, and cold bath-

ing,

ing, are likely means to prevent accumulations of bile. By attentions of this kind, the difeafe will feldom come on.

Let us now fee, what we are to hope from medicine, when there is an increafed fecretion of this fluid. Two plans offer themfelves, for the expulfion of bile; viz. to employ cathartics, and remove it by ftool; or emetics, and remove it by vomiting. When bile has been largely accumulated in the ftomach, and produced reaching, it is fometimes neceffary to affift the procefs by emetic means. This is for immediate relief. But unlefs in thefe circumftances, or preffed by fevere headachs, I imagine vomiting an improper plan. So far as my obfervations extend, I have uniformly found, that vomiting and naufea, in whatever manner excited or conducted, feemed always to increafe, and in fact to produce bile. I have, therefore, claffed thefe as exciting caufes.

In fea ficknefs, and in vomiting caufed by medicines; a great quantity of bile is often brought up, which is commonly fuppofed, the caufe of the previous ficknefs, though it is much oftner the effect of the naufea and reaching. One reafon reason would however induce me in particular cases, to employ full vomiting. I cannot conceive how such an immense flow of bile could assail the stomach, considering the situation of the common biliary duct, unless an obstruction either by spasm, or some other means, existed in the duodenum; below the entry of the duct. By this means the entry of the bile is prevented, and it is regurgitated into the stomach. To remove this obstruction and spasm, vomiting may be sometimes tried, and I think I have seen cases where it seemed to be useful in this way. But after all, I think the best plan of treatment, when accumulations have happened, is by laxatives and purgatives. Salts I have found to evacuate the bilious secretion more copiously, and to leave the patient more free from all its symptoms, than any other medicine. The only objection to their use, arises from the nausea they produce; this may in a great measure be obviated by dissolving the salts in simple cinnamon water, which I have often successfully practised. Laxatives, which teaze and irritate the stomach and bowels, do not seem calculated to evacuate the bile, for in fact, by their long continued irritation more is produced.

Perspiration

Perspiration and urine, are means of carrying the bile away. I have seen instances, where the urine tinged linen of a pretty deep yellow, where no jaundice existed; and the perspiration has often produced the same appearance in a slighter degree. In order to promote perspiration, I have sometimes prescribed, a few grains of James's powders, joined to a laxative; which being given over night, produced the double effect of increasing the discharge by the skin, and emptying the bowels. If the first dose of salts does not entirely free the patient from all bilious symptoms, a second dose must be prescribed; and the quantity of salts diminished, so as to obviate the nausea, which I have classed as an exciting cause.

Calomel, has been esteemed in India a most successful remedy, in all bilious complaints. I believe this opinion derived strength, from the great utility of it in hepatic disorders. Calomel, if given in a dose sufficient to purge, seldom performs this office without griping, and nausea; and if given in smaller quantities, it does not answer the purpose, and is very apt to produce its peculiar effects on the constitution. But the fact is, that calomel is seldom
prescribed

prescribed singly, it is commonly joined to cathartic extract or aloe, and aided by salts. It does not appear to me, to possess the amazing virtue ascribed to it in India; though I do not deny, that I have sometimes found it convenient and useful. The ease with which it can be taken in pills, has no doubt added to the character of this medicine, as the nausea which is sometimes the cause, and sometimes the effect of bile, renders it difficult to administer any bulky or disagreeable composition. Calomel pills, are certainly less nauseous and more commodious than salts; and the quantity may be so managed, as not to cause any great perturbation; but I am of opinion it never clears the stomach or intestines so effectually as salts. Many practitioners of India are of the same opinion. Their common method is to order a calomel pill at night, and a small dose of salts in the morning, and I believe this to be good practice. The calomel pill produces in the stomach and intestines, the purgative commotion; after which a very small dose of salts will procure a very free evacuation. Such evacuations must be repeated, till the patient is relieved from the symptoms we have already mentioned; till the languor and drowsiness are vanished. But calomel can only operate as a purgative,

purgative, with the difadvantage of having a rough operation, and of producing at times a falivation, a dangerous accident in warm climates.

Much may be expected, from attention to the occafional and exciting caufes in preventing a fit of the bile. General temperance, and moderate exercife, will greatly contribute to this end. When the fmalleft tendency, or the leaft fymptom appears, the patient ought to take a little caftor oil, an excellent and innocent laxative, or a fmall dofe of falts, and for fome days adopt a lower diet, and lefs exercife. If there is an habitual tendency to the production of bile, from the effect of heat alone, it requires minute attention to diet, and every circumftance already enumeratd, which conduces to excite the difeafe. Gentle riding and cold bathing, I have found in fuch circumftances to be highly ufeful.

The bilious habit is very difficult to cure, once it has eftablifhed itfelf. I have reafon to believe, that a flight mercurial courfe would be very ufeful in diminifhing the tendency to large fecretions in the liver. In feveral cafes, in the Eaft and Weft Indies, where the bilious habit

habit prevailed, I have had occafion to ufe mercury, for other difeafes, and remarked how foon the fyftem was fairly loaded or affected; that the bilious fymptoms abated and difappeared. This may be perhaps attributed to other circumftances, fuch as changes in the mode of living; but thefe were not fo fudden as to produce this revolution in the habit. Future experience muft decide, and enable us to determine it fully. Bile appears more or lefs combined, with all the fevers of India, and with the fevers of the Weft Indies; and I believe with the fevers of all warm climates; it is an attendant on our own autumnal fevers, and in various fhapes gives rife to diforders of the ftomach and bowels.

Savages, and rude nations, are in a great meafure exempted from this difeafe, particularly thofe, whofe religious inftitutions forbid wine and animal food; from this we are neceffarily led to conclude, that our refinement and luxury of diet, are the caufes of our being fo much troubled with this difeafe. It renders all the fevers in which it makes its appearance, more complex; but from the enumeration of its own particular fymptoms, many appearances may be explained, which
render

render the type of these diseases complicated. No disease is so often mentioned to the practitioner of a hot climate, as the bile; the inactivity and languor, which it produces, the loss of appetite, and dislike to all exertion, are no doubt very serious grounds of complaint. But unless the patient possesses more than ordinary fortitude and perseverance, in a plan of abstinence and restriction, there can be little done, without the assistance of an European climate; that is, without getting from the influence of a cause perpetually acting on the body. For medicines can afford temporary relief only, if the secretion is caused by the heat of the climate. Medicines are temporary powers, which cannot be used very often without danger; and they are opposed to the action of a power, which never ceases to operate, and always acts, with more or less vigour. In cases of this kind, where attention to diet, and the other means fail, and where the disease seems to result from the influence of heat alone, the patient ought to seek colder regions, as his only resource. Here he will in all probability recover, unless great obstructions have taken place. It may be a good general rule, to use purgative mineral waters, and to take a good deal of exercise on horseback

back or an open carriage; but if thefe fail, recourfe muft be had to mercury. When the influence of a cold climate itfelf, and the ufe of mineral waters, do not fucceed very foon after the patient's arrival, there is reafon to fufpect obftructions. In ordinary cafes, the change of climate alone is fufficient to produce every thing that is neceffary.

A great variety of ridiculous methods are ufed in warm climates, to prevent, or what they term, to cut the bile, in which confiderable confidence is placed; but which, of themfelves, have never appeared to me to poffefs any power. I have known much confidence placed in fwallowing a raw egg, beat up, fhell and all, in a mortar, and taken very early. This preparation might operate as a laxative, and certainly promote the evacuation of bile; but in any other mode, I cannot conceive it would have any effect whatever. If it remained in the ftomach for any time, from the vifcid nature of the white part, I fhould be inclined to fuppofe, that inftead of preventing, it would, by fupporting an irritation in the ftomach, caufe a larger fecretion of bile. At any rate, I cannot perceive any manner, in which it could be remarkably antibilious. The fhell

has

has too little calcarious fubftance to be of any great ufe, as an abforbent, and the contents of it, poffefs no chemical activity, to form new combinations or neutralize the bile. It is however a popular medicine, and as it does not do any perceivable harm, and flatters the hopes of the patient, I never forbade it.

Many fuppofe, that popular remedies of this kind have their origin, in a difcovery of real virtue in them: this may be fometimes the cafe, as accident unqueftionably has put us in poffeffion of valuable remedies; but in general, popular remedies are the offspring of fuperftition, or the cunning of quacks. It is however dangerous for the phyfician to interfere with them, as his oppofition is generally afcribed to ungenerous motives.

Another preventative of fome reputation, is the fwallowing of an orange, immediately on getting out of bed; this is a remedy much extolled. I confefs, I hold the fame opinion of its powers, with the preceding; it is believed, that its operation is laxative, and in this way, it may have a good effect.

But of all the remedies which are ufed to " cut the bile," fpruce beer has been held in the higheſt eſtimation, and I have heard many great drinkers of it declare, that it was impoſſible to be bilious, if only a ſufficient quantity of it was taken. I have tried this medicine myſelf, and taken a bottle of it, the moment I got out of bed; after which I rode, and I certainly found it, a very pleaſant and briſk laxative. With reſpect to the orange, I never eat one, when I had reaſon to ſuſpect the preſence of bile in my ſtomach, without feeling myſelf inſtantly ſick, and inclined to vomit; I do not know, whether others have experienced ſimilar effects.

Water creſſes, and lime water, have had their ſhare of reputation, as preventatives of bile; I cannot ſpeak of either, having never tried them. Water creſſes muſt however be a feeble means, and can only act, as a part of a ſyſtem of diet. Of lime water, I ſhould be inclined to think more favourably. It may ſtrengthen the fibres, of the ſtomach, and diminiſh morbid irritability, and thus contribute to diminiſh the ſecretions from the liver, by leſſening all irritation in its neighbourhood. No queſtions are more frequently put to phyſicians, and none more embarraſſing; than whether this or that diſh is bilious? Theſe

queſtions

questions are usually asked at table, where there is not much room for discussion. An answer must be given, and this answer goes abroad as a medical aphorism. It is a pity, that they do not recollect, that what may prove hurtful and bilious, in one stomach, may not have any bad effect whatever in another; and that it is excess in general, which renders any article of diet hurtful. But they believe, that certain substances dissolve themselves into bile in the stomach, as ices melt into their elements. It is right to inform them, that nothing of this kind happens, and that bile comes from the liver alone, without being formed in the stomach; from which however it may be derived in larger quantities, by distending the stomach with improper or indigestible food.

I cannot take my leave of Popular Remedies, without mentioning that spruce beer, acquired at one period great fame in St. Domingo as a sovereign medicine in the Yellow Fever. From the beginning, I gave no credit to the idle reports circulated in its favour. I could not reconcile to myself, that spruce beer, which had no perceptible action on the system, but as a laxative, could possibly change the course of so powerful a disease as the Remittent of St. Domingo. Overpowered however by reported inſtances of ſucceſs, and the converſation of every body; it became requiſite to give it a trial. I accordingly permitted ſome ſick ſailors, in various ſtages of the Remittent, to uſe this new remedy very freely. In ſome, it produced vomiting and ſickneſs; but in general, had its common effect as a laxative. But in no one inſtance whatever, did it appear in the leaſt degree to affect the courſe of the fever. Indeed our ſoldiers, could not have periſhed, if fortunately it had poſſeſſed any virtues, for they very freely indulged in this pleaſant beverage. Mr. WEIR made trials of it at the Mole, with the ſame liberal ſpirit, that always attends him. I had no opportunity of hearing the reſult.

The perſon who firſt promulgated the virtues of ſpruce beer, was one SMITH, the maſter of an American veſſel. He maintained that by its uſe only, he had preſerved his ſhip's company; and communicated his knowledge and doctrines, to the maſters of ſome Engliſh tranſports, who immediately became Pupils and Practitioners. The ſyſtem was ſimple and pleaſant, and peculiarly adapted to the palate of ſailors. It accordingly ſpread with
great

great rapidity among the shipping, who converted disease, into a social intercourse. But unfortunately numbers perished, either from too much or too little of the prescription.— SMITH however, who in the eagerness of system had not lost sight of his interest, accomplished pretty fully his views in practice. Besides a considerable reputation, for inspired knowledge, and being the founder of a new sect of physicians, he enjoyed the satisfaction of selling a considerable cargo of essence of spruce; which his new pupils greedily purchased at his own price. This imposture was ingenious, and has the advantage of being less prejudicial, than many other impositions on the Public.

I shall now speak a few words on what is termed,

THE PRICKLY HEAT.

The PRICKLY HEAT has been so termed from a sensation which attends this eruption, as if the skin was pricked in the several points which it occupies. The prickly heat begins to make its appearance how soon the perspiration has become general and conſtant, and ſpreads itſelf all over the body, beginning where the perſpiration is moſt profuſe. From the pricking ſenſation, which attends it, and an uneaſy itching, it becomes at times highly troubleſome, and a real diſeaſe.—There is a kind of ſenſibility ſpread over the whole ſkin, ſo that it will hardly bear the touch of the ſofteſt ſhirt, and renders every movement of the body painful and tormenting; but eſpecially ſo, before the commencement of a free perſpiration; juſt as the body attains full warmth.

This eruption, by the inhabitants of the Eaſt and Weſt Indies, is reckoned the beſt indication of a high and ſecure ſtate of health. They believe, that ſomething very injurious to the habit and conſtitution is now thrown on the ſurface, which had previouſly lurked in the inmoſt ſyſtem, and was inimical to the principles of life. They conſider no one in a ſtate of any ſecurity, until this eruption has made its full appearance; when they are perſuaded, he cannot ſuffer from any effect, the climate can produce.

The Prickly Heat, is undoubtedly a promiſing and ſalutary appearance, as it is the effect
of

of an eftablifhed and powerful perfpiration, which is certainly a difcharge of the higheft importance in a warm climate. It is indeed feldom, that any one, falls into a ftate of illnefs, or yields to the Remittent, where the perfpiration has been free, copious, and eftablifhed. The great mafs of fluids, is kept in a due ftate of coolnefs, the force of the blood is directed towards the furface, and a proportion of great importance, is eftablifhed between the bulk and expanfion of the fluids, and their containing veffels. It will be eafily conceived, that if the fkin becomes locked, and impervious, on the admiffion of morbid particles, that a chief fource of efcape, is barred againft them; through which in other fituations they might have paffed innoxious.

The Prickly Heat then, as a fign of free and copious perfpiration, is a very falutary and important eruption. But it is not critical, or does it confift of any injurious matter thrown on the furface of the body. For, we firft obferve, that it arifes with the commencement of perfpiration, and is increafed or diminifhed with the caufes, which increafe or diminifh perfpiration itfelf. Every one muft have remarked in a warm climate; that during the coolnefs of the morning, there is very little of this eruption vifible, nor is the fkin painful or uneafy; but when exercife, or the natural progrefs of the day, have directed the circulation more powerfully to the furface; a pricking painful fenfation immediately commences, the eruption begins as it were to emerge from the fkin, and becomes efflorefcent; until the actual commencement of perfpiration gives it, its full and complete appearance. It is obfervable too, that during the land winds, which lock the fkin very completely, and render the body hot and uneafy, the prickly heat is hardly apparent. It is not unufual on thefe occafions, to drink warm diluents, to reftore perfpiration; the moment it begins to make its appearance, the prickly heat begins alfo, and gives the firft warning of its approach. From this connection, and fubfequent appearance, uniformly prefent, between the perfpiration and prickly heat, the one always preceding the other, I think they may be claffed in the relation of caufe and effect. The prickly heat never made its appearance in any perfon, not fubject to very copious perfpiration, and copious perfpiration never continued for any time, without producing the prickly heat.

The

The caufes which increafe or diminifh perfpiration, likewife increafe or diminifh the prickly heat. But befides thefe relations, which mark them as caufe and effect, we fhall be able to account for the phænomenon, on this reafoning better than any other.

It will be difficult to fhow, that any matter injurious to the conftitution really exifted in the fyftem, previous to the appearance of the prickly heat. When fuch matter is any how introduced, it is feldom thrown on the furface, without the intervention of a febrile ftage, and after a certain progrefs, retires or fcales off, leaving the body perfectly free. But nothing of this kind is obfervable in the production or appearance of prickly heat; it comes on avowedly in a ftate of health, connected with a phænomenon the moft falutary that can happen; and without the prefence or affiftance of any febrile commotion. It has no ftated or precife period of exiftence, and does not retire or fcale off at any given time; but maintains its appearance, as long as the perfpiration is free and uniform; as long as the caufes which produce it operate. When thefe ceafe, or are diminifhed, the prickly heat difappears, or is confiderably leffened.

Let us attend to the manner, in which it would seem to be produced by the action of perspiration in the vessels of the skin. It is produced in the same manner, with eruptions which appear on the surface, in consequence of applying plaisters, such as Burgundy pitch, or the emplastrum roborans, that is, by exciting great and unusual action in the vessels, and supporting a continual perspiration. It is not improbable, that the perspiration in a warm climate is somewhat more acrid than in colder countries; it will be therefore more apt to irritate the mouths of the exhaling vessels on the surface, and at length to erode them, so as to produce the prickly heat. This eruption would seem to be produced then, in the following manner. The exhaling vessels on the surface, by the general direction of the circulation towards them, are made more irritable, which is still more increased by their perpetual action in pouring out the perspiration, which is itself an acrid saline matter, by which their orifices are eroded. These erosions pour out a lymphatic fluid, which incrusting on the skin, forms the eruption. The pricking sensation previous to the actual commencement of perspiration, would seem owing to the irritability of the vessels on the surface,

by

by which they are thrown into unufual action on the firſt approach of an uncommon quantity of blood towards them.

From this account of the Prickly Heat, which I believe to be juſt, from all I could ever obferve, it is evident, that there is nothing critical or dangerous in it ; that it is not matter thrown out on the furface, to relieve the body; and that it is in no other way falutary, than as it indicates, a free and copious perfpiration. From this account, it will alfo be evident, that there can be no danger from the retropulfion, or rather the retiring of prickly heat, except what may arife from the caufe that checked perfpiration. Many abfurd notions have been entertained on this fubject, which have really proved prejudicial to health. It has been very generally fuppofed, that the matter of prickly heat was highly injurious, and therefore, that the retiring of it into the body was extremely dangerous. It was remarked, that when it fuddenly retired or vaniſhed from the furface of the body, that difeafe ufually enfued, and the danger of it was afcribed to the matter of prickly heat again entering the circulation. The people who thus reafon do not recollect,
that

that whatever obstructs perspiration, or stops it entirely, whether cold, or the effects of fever, must also put an end to the prickly heat, which is only an effect of perspiration. The danger then, does not arise from the retropulsion of prickly heat, but from what caused the obstruction of perspiration, and perhaps from the obstruction itself. From false reasoning on this subject, many men are made extremely unhappy, who believe, that the least diminution of the prickly heat is dangerous, and attribute to this eruption every disease, or unpleasant sensation. Their life, is a cautious regimen, and their feelings, are alive to every change. I have known many persons, fall into a dangerous state of relaxation, because they would not continue the cold bath; for fear of beating in the prickly heat. Theories of this kind are dangerous, when they impede or destroy salutary habits.

Before I had an opportunity of attending to the prickly heat, I was biassed by the general prejudice, and avoided every thing, that I thought tended to repel it. I was afraid of the cold bath, and avoided it. On my arrival however in India, I became convinced, that the opinions entertained relative
to

to prickly heat were falfe and abfurd. I was determined to try an experiment on myfelf, and whilft my body was thickly encrufted with the prickly heat; I refolved to bathe. There was a large Tank of water in the neighbourhood of Diamond Harbour, and I chofe the morning, as being more cool and pleafant. I walked to it, without heating myfelf, and in the prefence of a number of gentlemen, who thought I was committing a very defperate action, plunged into the water, where I amufed myfelf for twenty minutes. I found no inconvenience from this practice, and repeated it every other morning, fometimes, in the middle of the day, and often in the evening. The prickly heat was diminifhed, becaufe the perfpiration received a temporary check, but with the return of that difcharge it returned alfo. But neither the fufpenfion of perfpiration, during the time of bathing, nor the confequent abfence of the prickly heat, which loft its efflorefcence, and feemed to retire, created the leaft degree of ill health; on the contrary, I found the cold bath attended with its ufual effects of increafing the vigour and hilarity of the fyftem. I could not for a long time prevail on any of the officers to follow my example,

from

from the dread of bad confequences. At length the feamen began to follow me, and ventured in. When they found it was not attended with any bad confequence; for all of them were covered over with prickly heat; they plunged in without referve, and often, when in a ftate of perfpiration, without ever feeling the leaft inconvenience. Such was the effect of prejudice, that if I had not in my own perfon, tried the experiment; I fhould probably continue to believe, that the prickly heat was a critical difcharge. The effect of fuch prejudices is often dangerous; we are deprived of a very falutary practice, and our views of difeafe are perverted: Thus a perfon, who believes, that the ftriking in, as it is termed, of the prickly heat, is attended with bad confequences, nay with imminent danger, lives in a ftate of perpetual anxiety; and in a warm climate, would avoid bathing; the moft falutary of all exercife. And when in a ftate of illnefs, if the prickly heat retires, inftead of attending to more important fymptoms, the chief attention would be directed to reftore this eruption, which has no connection with the difeafe, and is fuppreffed only in confequence of the diminution of perfpiration, perspiration, but in no other way adds to, or forms the disorder of the patient.

If I am right in affirming that the prickly heat is merely an effect of a very copious and continued perspiration; and depends entirely on that state, it will follow, that the precautions usually taken to guard against its retiring are useless; and that when it does retire, the danger does not arise from that circumstance, but from the action of a cause diminishing perspiration.

The prickly heat however, at times, rises to a height which constitutes disease. The patient, from the extreme irritation, is made uneasy, and some degree of feverishness is induced: in this situation, the physician is sometimes called for, and immediate relief expected. It will be evident, that no immediate relief can be reasonably expected; because the cure must depend on diminishing the perspiration, and averting from the surface the direction and force of the circulation. This cannot be done suddenly; some relief may be given by diminishing the cloathing; by laying aside flannel, and substituting cotton shirts; by avoiding diluent drinks, violent exercise, ercife, fmoking, and heated rooms, or meffes. But where the patient is very uneafy, and efpecially if he is full and plethoric; to the means already recommended, a blood-letting ought to be added, and afterwards feveral dofes of lenient phyfic. The patient fhould live on a lower diet, and take little exercife; he ought to drink lefs, and avoid every fituation, where he may be expofed to heat. After premifing thefe means, he fhould be directed to the cold bath, which will fo moderate it as to become very tolerable; if it is not entirely banifhed. But no degree of it forms any objection, to cold bathing among the troops; it is not attended with the fmalleft danger, and is a means of increafing the health and fpirits of the foldiers. I have feen this prejudice, prevent a number of men from bathing, in fituations, where they might have enjoyed this falutary and delightful luxury.

It may be ufeful too, to imprefs on the minds of practitioners and patients, this general truth; that no difeafe of any importance, ever had its origin from the ftriking in or retiring of the prickly heat. But that in cafes, where difeafe has occurred, and where the

the prickly heat retired; this has happened in confequence of a caufe, which deranged the fyftem in general, and diminifhed perfpiration. We are therefore, not to look to the retiring of the prickly heat in any important light; it is one of the fymptoms, but not a caufe of difeafe; and we are not to lofe time, in directing our efforts to reftore this eruption, which will of itfelf return with perfpiration. We are to attend to more important circumftances, and to direct our views to the general effects of the morbid caufe, inftead of combating one individual effect. In this manner danger may be prevented, and the difeafe brought to a happy termination. But whilft we were perfuaded, that the prickly heat poffeffed fome noxious quality, deftructive of life; our efforts were directed to throw it again on the furface. For this purpofe, heating means were employed, until the original difeafe was either exafperated into a more dangerous form, or fully eftablifhed itfelf in the fyftem.

I have remarked, that it was a very rare occurrence among the French, whofe habits are very different from ours. It is feldom or never feen, among the negroes, or the natives of the East Indies. Our systems are more heated, from our mode of life, and our perspiration more saline and acrid, than that of the French. They hardly know the prickly heat, whilst few of us escape it. It is however of consequence, to view it in its proper light, and not to suffer ourselves to be misled by an improper prejudice. One circumstance has chiefly contributed, to erroneous opinion relative to the prickly heat.—It was observed, on the application of cold to the body, that the prickly heat disappeared, and that some unpleasant symptoms occurred, which were attributed to the departure of this eruption. It was further remarked, that warm bathing, which usually brought back the eruption, afforded great relief, and removed the unpleasant sensation of the patient. The whole of the disease was of course attributed to the retiring of the prickly heat, and the recovery to its restoration. Now, the fact is, that cold, by impeding perspiration, and shutting up the skin, had produced the degree of illness which existed, and that this exhalation being for a time much diminished, or altogether absent; was the cause of the departure of prickly heat, and the warm bath, by relaxing the contraction on the surface, and restoring perspiration, restored

also

also the eruption, and removed the disease. But the whole of the disease consisted in a simple obstruction of the skin, and a diminution of perspiration, the departure of the prickly heat, being merely an accident, which had no share in producing symptoms, being itself an effect of circumstances, which had previously occurred.——

Having now finished, all the observations I had to offer, on Diseases in the West Indies, I shall proceed to the APPENDIX, and endeavour to show, that the reasoning I have employed on the treatment of Fever, has influenced physicians tacitly in the cure of most diseases.—

APPENDIX.

Reasoning of the Author confirmed by Practice.
 In INTERMITTENTS;
  FEVERS;
  ULCERS;
  LUES VENEREA;
  SMALL-POX.

IN the beginning of this book, I remarked that the practice, by definite indication, in Fevers, was not to be trusted, until greater light was thrown on the Proximate Cause: that our attempts must be directed, to effect speedy and powerful changes, to alter the whole condition of the body, to introduce new movements, and to impress on the system another mode of action. Let us examine with candour the treatment of diseases, and see how far this practice, though not acknowledged in terms,

terms, has prevailed.—And firſt let us proceed to

INTERMITTENTS.

It will probably for ever remain a ſecret, how theſe aſſume their peculiar and diſtinguiſhing types. We cannot form any theory relative to this point, that can afford the leaſt ſatisfaction to a juſt thinker. Leaving this inveſtigation, to ſome fortunate genius; I ſhall attend to the different methods of cure, which have occaſionally proved ſuccefsful.

When we attempt to preſcribe in Intermittents, from a knowledge of their proximate cauſe, we find ourſelves very ſoon in obſcurity. The remote cauſes, are indeed, pretty well aſcertained, and the ſituations in which Intermittents uſually ariſe, are likewiſe well-known. But of the preciſe condition, which conſtitutes the proximate cauſe, we are entirely ignorant. From a difference in the condition of the ſyſtem, or the modification in the remote cauſes, marſh miaſmata ſometimes create Intermittents, ſometimes Remittents or Dyſentery. We remark, that an exposure to these miasmata, after a certain period, produces a peculiar mode of acting in the system, which brings on cold shivering, heat and sweating; and disposes these phænomena to disappear and return, in a certain periodical manner. It is acknowledged, that the proximate cause is unknown, but we know, that its mode of acting must be changed, or itself banished, before any thing can be done for the effectual relief of the patient.

When we observe one cause, producing various effects in different bodies, it follows, that the cause itself is modified; or the bodies to which it is applied, determine its action in a particular manner. It is remarkable, that a contagion, evidently the same, when applied to different bodies, produces effects so very various in each. It is not therefore improbable, that the types of Remittents and Intermittents, result from the habit; and not from any specific variety in the morbific miasmata. It were much to be wished, that we could ascertain the precise state, which determines the type; but this I fear is not to be attained. Our prescriptions in Intermittents, are founded chiefly

chiefly on experience, gained from cafual obfervation, or accident; for we cannot found indications, on a knowledge of the proximate caufe. It is true, that phyficians have fuppofed it to confift chiefly, in an atonic ftate of the extreme veffels, and have afcribed the good effects of bark to its tonic powers. But it is not by any means evident, that this atonic ftate exifts in all cafes, nor is it at all clear, that the bark effects a cure, by communicating tone. If bark operated in this manner, other tonics would produce equal good effects, in proportion to their powers; but this does not happen. Befides, it will prefently appear, that other means effect cures in Intermittents, which do not in fact poffefs any tonic powers. The truth feems to be, that a fecret change is produced in the morbid action, which at laft ceafes, and the ufual movements are reftored to the fyftem. We direct our views to create a change, in the general action of the fyftem, fo as to difpoffefs the peculiar modes of the morbid action. Such a change, Peruvian bark is known frequently to effect, and experience fupports its ufe. But other means have been alfo employed with

with fuccefs, in changing the morbid action, and reftoring health to the fyftem. Some of thefe prevent only the return of one paroxyfm, whilft others are attended with more permanent benefit. An emetic given at the time a paroxyfm is approaching, fometimes entirely prevents it; and the ingenious LIND, has fhortened the duration of a fit, by the ufe of opium. Now thefe means, feparately examined, are in themfelves different; and produce different effects. They are however calculated to effect a change, which banifhes the morbid action.—

If it be faid, that bark cures Intermittents by giving tone; we may remark, that the operation of emetics is not tonic, nor has laudanum, any ftriking power of this kind. When fpiders, or other difgufting animals, are given to aguifh patients, the practice is ftrictly founded on the plan of changing morbid action without definite indication. Horror is excited, and the fyftem is under the influence of a powerful change, which fuperfedes the agency of the morbid caufe. I cannot in any other manner, account for the good effects, which have fometimes arifen from fwallowing fuch animals. To the fame account may be placed the benefit of of dashing water suddenly on the patient; which has often shortened the duration, and meliorated the whole of a paroxysm. Exercise, and the effects of interesting intelligence, fall into the same class. We cannot account for their effects in any other manner.

Steel, is on many occasions a powerful and useful tonic; but it possesses little virtue, in curing Intermittents. The kind of change it produces, does not seem calculated to overcome the morbid action; an argument, that something more than want of tone constitutes the proximate cause.

I have in numerous instances cured Intermittents, in India, and at sea, by the use of calomel, after bark had entirely failed. Now mercury, has never been supposed to give or to produce tone; on the contrary, its action is commonly attended with debility; unless in cases, where it removes a more powerful enemy to the constitution than itself.— Thus, it restores vigour, to constitutions worn out by *lues venerea,* or weakened by the violence of hepatitis. It acts however in the cure of Intermittents, not by
any

any specific power, directed to any individual effect of the proximate cause; but by causing a change, in the general movements of the habit, the morbid action is at length banished.

A solution of arsenic, has been employed very successfully in the treatment of Intermittents; and seems to produce its effects in a similar manner. And it would appear, that bark itself, as Dr. JACKSON remarks, produces its effects, not by any specific power, by which it would in all cases act successfully, but by introducing gradually into the habit counter movements. Methods the most various, have sometimes produced the most happy effects, so that we must conclude, that these arose, merely from inducing changes.

In the remarks now offered on the manner of treating Intermittents, it has been observed, that various plans frequently succeed in the prevention of paroxysms, which do not appear to be directly calculated, to obviate the proximate cause. The indications are not formed on any definite knowledge of this subject. We merely effect a
cure

cure by producing a change. Bark, which frequently fucceeds, is by no means a remedy always to be relied on. I have met with many agues of the tertian and quotidian periods, which refifted the ufe of bark in the largeft dofes, and after a long continued ufe. The Intermittents of Bengal, particularly furnifhed thefe inftances; thofe of China, yielded to the common treatment. In cafes of this kind, even where great debility prevailed, I ufed mercury, and was never difappointed in my views. Soon after the patient commenced this courfe, the paroxyfms became more mild, continued for a fhorter time, and in all refpects changed their violent procedure. That is, there was fomething in the action of the proximate caufe, which the feeble effects of the bark could not banifh, but which gave way to the more potent operation of mercury.

The native practitioners in India, from a kind of inftinctive knowledge, for they are very illiterate, purfue the fcheme of introducing changes in the fyftem. When at Calcutta, I happened to converfe with one of them, on their method of curing the Intermittent, which not unfrequently attacked the

the inhabitants. He informed me, that they purſued a variety of ſchemes; that they ſometimes poured buckets of cold water, on the patient, during the cold fit, and afterwards wrapped him up in warm coverings in bed; by this means he alleged, that the cold fit was ſhortened, and the hot brought on. He ſaid that the ſweat flowed more freely, and that on the whole, the paroxyſms were ſhorter and milder from this practice. He ſhowed me ſome powders, which had an aromatic ſmell like caſſia, which he had prepared from dried herbs; but he would not produce the herbs themſelves, nor inform me, where to procure them. He ſaid, they effected cures in a very ſhort time, among their own people; but that the blood of Europeans, being more hot and inflammatory, required more powerful medicines. He indeed produced a nut, of an olive colour, covered with an elaſtic, flexible huſk, about the ſize of an almond, it contained a bitter taſted kernel, with a ſmall degree of aromatic flavour. The method of uſing this medicine, according to him, was to bruiſe the kernel with a few grains of common pepper, and forming the maſs into pills, to adminiſter them frequently till the

the paroxyſms, at length gave way. He called the nut; kút ka léeja, or lota kâ pūl. I ſupplied myſelf with large quantities of the nut, and tried it, in many caſes, with excellent effect, but could not truſt the cure entirely to their uſe. It was a powerful and good bitter, it warmed the patient; and created a grateful and pleaſing ſenſation in the ſtomach.

From what has been ſaid on the variety of methods, in which Intermittents may be cured, it is evident, we do not practiſe on any definite indication; we either try changes, which experience has already ſanctioned, or ſeize an analogy; and exhibit ſuch remedies, as in other caſes are known to produce powerful effects. The treatment of continued fevers, is founded on the ſame principles.

Of CONTINUED FEVERS.

CONTINUED FEVERS have been an opprobrium to phyſicians in all the ages of medicine. The Ancients have thrown little light on the ſubject, nor have the Moderns been much more ſucceſsful. Ingenious
ſyſtems

fyftems have been offered, and rejected.—
Theory, which though fpeculative, often
influences the phyfician, feldom had vigour
enough to change practice, becaufe it was
commonly rather an effufion of ingenuity,
than an induction from juft reafoning. On
many occafions, theory and practice have
been at variance; and in general there was
little union, between fpeculation and experience. Cures occurred under the moft op-
pofite modes of treatment, and the confidence, which this cafual fuccefs infpired
gave currency to particular remedies.

The Ancients, in their cure of Fevers,
for a long period, continued the ftrenuous
imitators of their predeceffors, without af-
piring to truth or novelty. The remote
caufes of continued fevers, are undoubtedly
obfcure; but above all, the proximate caufe,
or what more immediately exhibits the morbid phænomena, has eluded every refearch.
It is ufelefs to repeat the various conjectures which at different periods occupied
the medical world; it is fufficient to remark,
that none of them have ever led to a decifive, or certain plan of cure.

Fevers,

Fevers, of the continued form, assume, at times, the type of Intermittents; that is, there appears a certain degree of abatement in the symptoms, at stated times; but they again resume their wonted course, in periods corresponding with the returns of tertian paroxysms. This is the most frequent type at least. But whether this depends, on something in the constitution, which determines the return of the fever, or on the operation of powerful causes, is not known. But there are continued fevers, in which no perceptible abatement is evident, and they run through their whole course, without suffering any visible or apparent change in the severity of the symptoms. The operation of the proximate cause, in continued fevers, is steady and powerful; and seems as yet, to have bid defiance, to all the suggestions of theory, or the dreams of credulity. On a survey of the practice, which has obtained in continued fevers, through all the æras of medicine; I confess, that in my mind, it has been uniformly too feeble. The practice of the Indians in America, appears to me to possess more vigour, and to be more likely to do good, than all the systems as yet promulgated by the schools of

physic

physic. Whatever the proximate cause may be, which produces the morbid action, and exhibits the phænomena of fever, it seems to be tenacious, and to keep possession with wonderful perseverance. Such a cause is not easily moved; powers which produce slight changes are not likely to affect it; bold and decisive practice must be adopted before we can do any thing; and as we cannot pitch on the weak part, for the play of our engines, let the whole system be stormed at once, and the disease banished by a powerful invasion. From the want of this energy in practice, and the influence of idle theories, the treatment of fever, has been feebly conducted. Nor has it ever been clearly proved; although affirmed by credulous or dishonest practitioners, that the course of a fever, was really cut short by these tame operations. It is at least probable, that the proximate causes of disease are only to be removed, by the introduction of counter movements, which effect a general difference, in the action of the whole system, or its parts. Now we see, that the practice in fever for a period of two thousand years, had not introduced changes sufficiently powerful, to remove with certainty the operations

tions of the proximate cause. The morbid action, we must conclude, is very powerful, it does not seem on any occasion to give way, to feeble opposition. Slight attempts avail nothing; in such cases, there is room for innovation. If we are persuaded, that sudden changes, or revolutions, are proper, let them possess energy to effect their purpose. The prejudices of mankind, and the fears of practitioners, oppose this general scheme of treatment, but I have no doubt, but the bold physician, will be crowned with frequent, and unexpected success.

I have often seen remarkable effects from sudden changes, applied in such a manner, as to alter the whole circumstances of the habit. When we see a fever obstinately resisting ordinary means, and sweeping without distinction, the toiling race of man; is it not then incumbent on us to vary our means, and increase the chance of success by multiplied efforts? In such disasters any new plan, can hardly be less successful than the old one; and experiment may at length put us in possession of a better method. All our present knowledge must have at first arisen from chance trials. It is from experience alone, that principles can be deduced, or enlarged,
that

that hints can be extracted, which prosecuted by further enquiry, may become the basis of systems. Investigation, is slow and laborious, we generalize and extend from small beginnings; but the philosopher is rewarded by the discovery of truth, by conferring on mankind durable benefits.

From the great mass of casual experience a selection is made, which may serve to enlighten posterity. The method of practice, by inducing a revolution or change in the habit and constitution, and thus banishing morbid action, extends our views, and gives a scope to the physician, which he could not otherwise attain. The doctrine applies to a number of diseases, and gives a new foundation to practice, when indication wholly fails. I shall show in a few instances, its direct application to other disorders, where the practice by indication could have no place; as the proximate cause was wholly unknown.

In the remarkable history related by KAU BOERHAAVE, of the powers of irritation and sympathy; we see a wonderful instance of the force of terror in changing a morbid action,

action, that had established itself fully in the system. The striking figure of BOERHAAVE, his solemn, awful deportment, his determined manner, impressed fear, and excited movements in the system, which banished and conquered the influence of the morbid cause. This memorable history confirms very strongly the reasoning on morbid action, and the manner of changing it, by sudden and powerful means; for in no other way could the mere appearance of BOERHAAVE produce any effect. The surprise however, and impressions of fear, by altering the movements of the system, banished the morbid, and restored the healthy movements of the constitution.

There are numerous cases on record, where sudden and powerful revolutions, have wonderfully affected the body. In Lord ANSON's voyage, it is related, that on one occasion great numbers were prostrated by the scurvy. A ship however came suddenly in sight, supposed to be an enemy; the men were roused, and became anxious and eager for battle. The appearances of disease greatly abated, and they seemed, as it were, to have at once recovered.

vered. They soon afterwards relapsed. The effects of surprise and novelty, banish an obstinate hiccup.

Now there is nothing in these cases, particularly directed against any individual effect of the morbid cause; the whole action of the system is suddenly changed, and in this manner the morbid action is banished. It has been remarked by almost every one, who has ever followed an army, that men languish and become sickly in easy quarters, but recover very rapidly when their powers are awakened, by the approach of danger, or the expectation of an enemy.

I shall now record an instance of the astonishing effects of sudden changes, or powerful movements in the system. When I was proceeding to Bombay, in the Middlesex East Indiaman, a continued fever broke out on board, which attacked great numbers; though very few died. We touched at the Cape of Good Hope for refreshments, and proceeded on our passage. The fever still continued to affect the seamen, and they lingered under it for weeks. When we came however to lat. 36° 19′ S.

eight

eight or ten of the people had very unfavourable symptoms. Next day a violent gale arose, with a tremendous, tumultuous sea, agitating our ship with rapid and uncommon motion. It exceeded in violence, all the storms and tempests, the oldest men amongst us could remember. It was wholly out of my power, for two days, to visit the sick, or give them any assistance. When I ventured amongst them, on the third day, I expected to have found several dead, and the others much worse. They had hardly received any nourishment, and little attention of any kind, during the continuance of the gale. But how great was my astonishment to find, when I visited them, that they were all free of fever, and complained only of debility. The course of the fever had been entirely stopped. No one will here argue, that there was any prescription, founded on direct precise indication; the morbid action, which previously existed, was changed by very powerful movements in the system. The dreadful agitation of the ship, appears the chief agent, which by a continuance of three days could not fail to bring on important changes. Fear, hope, and a variety of strong emotions, must have
alternately

alternately prevailed. Sudden changes then, have in many instances produced cures, by altering at once, the whole given circumstances, and condition of the body, and by introducing a set of movements totally different, until the system at length adopts its usual and salutary action. This is in no way founded, on partial indication, or any individual effect of the morbid power. I am not acquainted with any indication in continued fever, that would lead me, to be in any measure confident of success. We sometimes obviate pressing symptoms, and remove stimuli, which might support irritation, and we endeavour to support the vigour of the vital powers, until some change may happen; and this comprises all our knowledge in the treatment of fever. In such cases, I should be strongly inclined to pursue bolder means, and endeavour to change the phænomena. It is however evident, that most physicians have aimed, in the treatment of continued fever, to bring on sudden changes in the habit, without consulting lesser indications. Whilst the doctrines of the venerable CULLEN prevailed, the removal of spasm, and the

giving

giving of tone, were the great views of practitioners.

OF ULCERS.

THE late ingenious Mr. JOHN HUNTER has furnished many curious remarks on morbid action. Let us take a short view of the treatment of ULCERS; and see how far it is founded on our principles. Wounds, in whatever manner produced, from certain unknown causes, in some instances become foul, and unhealthy ulcers. The matter secreted in them is either thin, acrid, or sanious. The peculiar modification of action, which then exists in the vessels, disposes them to this untoward discharge. In these circumstances, a cure is not readily effected; the ulcer proceeds to acquire a worse appearance, and the aid of physic becomes requisite, to give it a better aspect. In these cases, before we can do any thing of the least advantage to the patient, a change must be effected in the mode of action, by which the vessels must be disposed to another modification,

more

more favourable to the production of good pus. Many authors have been convinced, that this change has been produced, by a certain management of heat, on the secretion itself; independent totally of the action of the veffels. They became more firmly perfuaded in this belief, fince the publication of Sir JOHN PRINGLE; where fome experiments on this fubject are detailed. The medical character of Sir JOHN PRINGLE, ftands defervedly high, but he has been mifled by the circumftances, on which he grounded his theory. In thefe experiments, ferum was expofed to a regulated heat, and after fome time, a whitifh coagulum, was difcovered at the bottom of the crucible, with a fœtid difagreeable fmell. From thefe appearances, it was concluded, that the mode in which pus was formed, had been difcovered, and that the whole procefs depended on modified heat applied to extravafated ferum. Mr. BELL, of Edinburgh, in his treatife on ULCERS, a work of confiderable ufe and merit, has adopted this reafoning. To me, the matter appears altogether different. It would feem that the veffels of an inflamed tumour themfelves, communicated to the ferum, by

a pecu-

a peculiar action, the power or capability of becoming pus. And, that the nature of the discharge from any ulcer depended, not on a regulated heat, but on the peculiar action, that may at the time exist, in the vessels of the part, or in the system at large. Thus in a foul ulcer, it happens that no impression can be made on the discharge, by any regulated heat, in any form whatever, yet the exhibition of the bark, produces astonishing effects in a short time. No one, I presume, will argue, that bark in this instance operated on the serum itself, which may be supposed out of its reach, and extravasated in the cavity of the ulcer. The favourable change appears evidently to result, from the action of the bark on the vessels, and system in general, and changing the peculiar state which gave rise to the untoward discharge; by changing this modification of action, the vessels are enabled to endow the serum, with the capability of becoming pus. We remark besides, that tumours undergo an intermediate stage, before pus is produced; notwithstanding the application of poultices or fomentations. If heat alone could convert the serum into pus, this intermediate stage of inflammation

tion would by no means be neceffary. But the difpofition or capability of becoming pus is communicated to the ferum by the veffels, and not by any modification of heat, or any action of it, on the ferum, after it is once fairly fecreted. Venereal fores affume commonly the moft unfavourable afpect, and the difcharge is acrid, thin, and offenfive, until mercury be given. In vain will poultices be applied, or bark adminiftered, till this medicine has been given. That is, the peculiar action, which exifts in venereal ulcers, and conftitutes their effential nature is not dependent on the ftate of the ferum, but on that of the veffels, which is only to be changed by mercury. We may remark too, that ferum is often extravafated in other cavities, and expofed to confiderable heat, without becoming pus. The peculiar action which creates pus, not being prefent in the veffels, when fuch extravafation happened, the neceffary difpofition to form it was not beftowed.

The utility of regulated heat, in the form of poultices and fomentations, is confirmed by experience, but certainly does not produce

duce its effects, by acting on the extravasated serum. The collections of water in dropsy, bear a strong analogy to serum, yet we do not find that heat converts them into pus; the reason is, that the vessels did not bestow that peculiar something, which is necessary to this process *.

The admixture of solids, the dissolution of the vessels, and the addition of blood, do not afford any satisfactory explanation of the phænomenon. The various secretions of the glands, are performed by a peculiar modified action; for although the elements of bile, urine, and semen, may by analysis be discovered in the blood, yet no one has ever detected them in their appropriate peculiar form; nor are they ever seen, before the vessels have performed their peculiar act, by which these secretions are produced. The blood itself seems to derive from the action of the vessels, its colour and determinate nature. They exert upon it a peculiar action, and successive changes complete the process. The stomach

* Vide Note I.

separates

separates the nutritious parts of the aliment, which change into new qualities, by the admixture of bile and the pancreatic juice. The lacteals perform their part, and further changes are effected in the progress of the chyle, through the thoracic duct. Till at length, the grand operation, which finally determines the essential nature of blood, is performed in the lungs, heart, and arteries. From this seeming uniform mass, are derived various fluids, by the action of particular organs. Frequent changes happen in these secretions, when general health is by no means impaired; thus the colour of the urine, and the consistence of bile, are hourly varying, from a change in the mode of action, in the glands which perform these secretions. To a change in the action of the kidneys, may be ascribed the foundation of gravel stones. They are not always the effect of a nucleus, casually existing in the bladder. The mode of action in the secretory vessels, disposes particles, to separate themselves from the urine; and thus produces, if I may so speak, the gravellish tendency. Baron HALLER informs us, that the presence of females, promotes the seminal secretion, and excites uneasy

sensations

fenfations in the glands themfelves. This unqueftionably proves, that a peculiar action is produced; and the immenfe flow of pale urine in hyfteric females, fhows, what increafed or varied action may perform.

I have adduced thefe inftances, in fupport of the action of the folids, which appears to be the chief agent in producing the various animal phænomena; but practical benefits refult from correct reafoning. Thus the theory of Sir JOHN PRINGLE relative to pus, may on feveral occafions, lead to an inert or improper practice. It would for inftance, lead to local applications, in cafes, where the fyftem at large ought to be acted on; and we fhould be lofing time in feeble efforts, inftead of purfuing a manly vigorous treatment. Thus far, falfe theories are dangerous, and merit ftrict attention. Whenever our practice is the refult of theory, it becomes us very narrowly to examine it. Phyficians, and among others Sir JOHN PRINGLE, have been wonderfully deceived by the application of their inductions from experiments; the very principle of fuch experiments being erroneous. I mean experiments

periments made on various substances out of the body; or on dead animal flesh. It is unphilosophical to reason in this manner, or to expect any useful induction from any number of experiments, made on substances, whose qualities with respect to each other, are so remarkably opposite. We cannot hope for useful discovery, in this mode of investigation. The animal and dead fibre, differ so widely, as not to admit, almost, any common analogy. Experiments of this kind, may be useful to commerce, and enable victuallers and commissaries to preserve provisions; but cannot be of the least utility in medicine. Camphor has been found in these trials, to resist the putrefaction of animal fibres, and has been since employed as an antiseptic in fevers; but if it possesses any power of this sort, it arises from a very different source, from that to which we ascribe a preservative quality, with respect to dead flesh. Let it be observed, that common culinary salt, in a remarkable manner, resists the putrefaction of dead flesh. Yet no one has thought of prescribing it in scurvy, which is often attributed to its use.

Upon

Upon the whole there is no credit due to experiments made on any ſubject, except the living human body. In ſome very broad analogies, from the brute creation, we may receive principles, though I am inclined to believe, that we have been led into errors, from proſecuting too earneſtly this manner of reaſoning. The conſtitution of brutes is very different from ours, and we are ſubject to much fallacy, becauſe we muſt be ſatisfied with what we can obſerve, without the benefit of interrogation.

Upon the whole, it would appear, that in INTERMITTENTS, CONTINUED FEVERS, and ULCERS, we effect cures, not by directing our efforts to a preciſe known proximate cauſe, but to effect a general change in the ſyſtem, and by altering all the circumſtances baniſh at length the morbid action.

Let us attend to the progreſs and cure of

LUES VENEREA,

And ſee how it accords with theſe principles.

The

The matter of Lues Venerea, when applied to certain parts of the body, produces a peculiar morbid action, which characterifes the difeafe, and which yields to the changes, experience has taught us, mercury can produce. It evidently arifes from contagion applied to the body, in an-active ftate; and this contagion generates and produces the difeafe. When we examine the fubject more narrowly, a regular progrefs is obferved in all the phænomena. The contagion, foon after its application to the fyftem, finds a neft, where by a power unknown to us, it feduces the veffels from their ufual mode of acting, and produces in them a difpofition to fecrete the venereal virus. This peculiar modification of action in thefe veffels, whatever it be, forms the proximate caufe; and exhibits the morbid phænomena. In this manner chancres, and other venereal appearances, are produced. When a large portion of the living fyftem is under the influence of this action; the quantity of the virus is increafed, till at length, there appears a general tendency to adopt the new impreffion; when the habit may be faid not improperly to be venereal. This is the manner of its progrefs. Experience has fortunately taught us, that this morbid action, so formidable to youth and pleasure, and so dangerous to the procreative faculties, yields to the action of mercury. The object of the physician is to change the morbid action, introduced by the venereal virus, and to substitute another set of movements, so as to bring the system back, to its ancient laws, and common action. But we cannot perceive the peculiar mode in which the virus acts, so as to produce its phænomena, nor do we know precisely the manner in which mercury brings the system to health, and cures the disease.

Experiments have been made to ascertain, whether there was any chemical affinity between the venereal virus and mercury, by which they might be disposed to unite, and form an inactive neutral. These experiments, like all others made out of the body, gave no satisfaction. For, if it had been proved, that such attraction actually existed; little or no light, would be thrown on the subject. It would not follow by any means, that this combination took place in the living system. We have numerous instances in practice, where the local application

cation of mercury is daily made, without producing any good effect, until the fyftem in general was affected; that is, till another action banifhed the venereal movements. Nor does mercury itfelf feem to poffefs a perfect fpecific power, by which, in all cafes, it would effect a cure. There are circumftances of the conftitution, which by affecting either the morbid action itfelf, or the operation of mercury, prevents the habit from recovering. Too much vigour, or too great debility, impede and retard the fuccefsful operation of mercury. I have feen inftances in fcrophulous habits, where the utmoft difficulty occurred, of making mercury at all ufeful; and where there was fomething in the morbid action, which refifted all oppofition, and proceeded to the full deftruction of the conftitution.

Lues venerea, is the confequence then, of a contagion perverting the ufual movements of the fyftem, and feducing the veffels of the part to which it is applied into a new action, which induces them to fecrete a matter fimilar to that which excited the commotion in the habit. And mercury is another power, which by affecting the fyftem, in a more vigorous manner, banishes the action of the venereal virus, and brings back the system to its usual obedience, to its common laws. But the definite manner, in which it operates, we do not know, nor do we know the proximate cause. No one will pretend to say, whether the action of the vessels, where venereal ulcers arise, be quick, or slow, or oscillatory. There is a secret modification of action in the proximate cause, which the wisest of us have not been able to ascertain; nor the manner in which mercury banishes it.

We observe, from experience, a number of circumstances which influence the favourable, or unfavourable effects of mercury; these as matters of fact, for we cannot account for them, guide our practice, and influence our prescriptions; and this is all the knowledge we really possess. There are undoubted proofs, that the state of the body changes the whole phænomena of a disease, so as to give the effects of one cause a total different aspect. Thus the matter of lues venerea applied to glandular or secreting surfaces, brings on a morbid discharge, with many other symptoms very different
from

from the venereal chancre. And yet no doubt can be entertained, that the matter which produces both, is entirely the same. I know, that other opinions have been advanced on this subject, which may be very ingenious, but are certainly very remote from the truth; as discovered by experiment or analogy.

Let us next attend to the phænomena of a dreadful disease, the

SMALL-POX.

The matter of the small-pox, when introduced, forms, like the venereal virus, a nidus for itself, and there exerts its peculiar powers, by reducing the vessels of that particular spot into a new action, by which they are induced to secrete a fluid, every way similar to the original contagion. This forms a kind of magazine, from which particles of matter are supposed to be detached into the blood, these by some unknown law of the system, are directed to the surface where they form nests to themselves, and undergo a similar process, to the matter at firſt lodged. Each ſpot has its intermediate ſtage of inflammation, or its proximate cauſe, or ſecret action, by which the matter of ſmall-pox is finally and completely produced.

The fever of the ſmall-pox, which precedes the eruption, appears to me, to be the revolt of the ſyſtem from its common laws, to adopt the new or variolous action. It is doubtful perhaps, whether particles in a ripe formed ſtate are really detached to the ſurface of the body; it is more probable, that the variolous matter acts from the ſpot in which it was at firſt depoſited by the intervention of the nervous ſyſtem or ſympathy, and in this manner ſeduces the veſſels of the ſkin, where if a tendency to inflammation happens to be preſent, the variolous action will become general, and a large quantity of the matter will be produced. This reaſoning is countenanced, by obſerving, that no matter, in a formed ſtate, is diſcovered in the early variolous puſtule, which is a hard inflamed ſpot, that produces, after a certain period, the matter of ſmall-pox. The inflammation or peculiar action ſeems abſolutely neceſſary to the production of the variolous matter. If this

is

is not true, the future difeafe, ought always to bear fome proportion to the quantity of matter at firſt introduced, or afterwards generated in the nidus; but it is known, that no fuch proportion has ever taken place. Nor would the puſtules ever require the intermediate procefs of inflammation, but would appear at once full of mature and ripe fmall-pox. So that it is not quite clear, that matter is pofitively difperfed in a formed ſtate over the body. The proximate caufe of the varioli, then, is a fecret modified action, which induces the veſſels to adopt new movements, and feparate the variolous matter. The veſſels under this influence are, for the time made glandular, and every fpot may be juſtly confidered as a gland fecreting a peculiar fluid.

We are evidently ignorant of the proximate caufe of thefe phænomena, fo as to prefcribe for it, and baniſh it. Experience has indeed ſhewn us many circumſtances, connected with the favourable or unfavourable progrefs of the difeafe, and thefe are embodied into regulations for our conduct in the treatment of it. But we do not proceed further, we have no definite or decifive

cifive indication; we cannot prevent the difeafe, but by flying it, nor can we with certainty amend its fymptoms, or pofitively cure it. The veffels on the furface are the feats of the difeafe; on their condition much muft depend, and on the kind of action they adopt. The application of cold, in the manner of the celebrated Baron DIMSDALE, feems to put them in a very favourable condition for a mild fecretion. The inflammatory ftage is moderated; on which fo much depends in all difeafes, where matter is formed; and the whole procefs is made more mild and gentle. The fpecies of action, which would produce a malignant kind, is thus changed, and a modification introduced, which conducts the difeafe to a happy iffue.

I have no doubt but fome important improvements may be yet made in the management of this formidable malady. The late ingenious and learned CULLEN, when treating of the fmall pox in his fyftem, rightly imagined, that a peculiar ftate of the veffels on the furface, regulated the future events of the difeafe. Thefe veffels certainly appear to be the principal agents in the

the production of the variolous matter, and therefore must influence the issue. It is singular, that the variolous matter, having been once general in the system, cannot be again reproduced, by applying the contagion; this is one of the mysteries in the animal œconomy, which we may never be able fully to explain. There may however be a chance of discovering a remedy, which, like mercury in the lues venerea, may counteract the proximate cause, without going through its usual revolutions. I think there cannot be a doubt, that the variolous disease is produced *, in the manner we have been stating; and that improvement in the manner of treating many diseases may result from observing minutely, the laws which regulate morbid action.

We have thus shewn, that in many diseases, although we evidently mark a peculiar morbid action, constituting their proximate cause, yet we are not sufficiently intimate with its peculiar mode, so as to prescribe for it, or change it; and therefore, that our practice, when directed to some

* Vide Note II.

of its effects, becomes feeble, as the cause continues its operation. But when we attempt to change the whole given circumstances of the body, and introduce sudden and considerable changes, we have a chance of banishing the disease from its strong holds, and when its morbid action ceases, the system naturally adopts its ancient laws, and usual movements.

NOTES.

NOTE I.

THERE are some circumstances in VENEREAL ULCERS, which I could not so properly blend in the discussion of that subject; but which may be added here. It has been remarked, that the venereal disease, does not seem to depend, on the nature of the matter which produced it; that is, on the greater or lesser vigour of the virus, but on the state of the system to which it is applied. This I believe in general to be found true.

I shall just mention one condition of a venereal ulcer, in which it may be possible, that the virus, is much below its usual standard of vigour, and in this state, may, from want of strength in the virus, bring on a milder disease. This I confess is mere conjecture, which I have not been hitherto able, precisely to ascertain from experiment.

experiment. This situation occurs at the critical period, in which the venereal action is about to cease; and before a perfect cure is effected. It occurs in gonorrhœas, when about to degenerate into gleets; when the affected vessels, are under that compound influence, where the secreted fluids, are neither morbid, nor perfectly healthy. Such a state may really be supposed to exist, near the cessation of gonorrhœas, and when venereal ulcers, losing their characteristic appearance, begin to put on the aspect of simple sores. It may be difficult to conceive that the vessels can be under the influence of a double action at the same time; but in a state of disease, they always are, for the healthful movements of the vessels are never wholly destroyed, till death. Projectile bodies are under the influence of two powers, the projectile itself; and the power of gravitation, the result of which is the parabolic curve; it may be thus in the human body, two actions may exist, which at length terminate in the usual movements of the system.

It has been remarked in the East Indies, and I can bear testimony to the fact, that
the

the venereal diseases of Bombay, are more violent in their progress and effects, than in any other quarter of that country; Dr. KAY, of St. Helena, an experienced and acute observer, remarked, that most of the desperate cases he had seen, on their return to Europe, had come from Bombay. To what is this difference owing? is it to the nature of the virus? or in something peculiar induced in the constitution, by the climate of that island?

NOTE II.

It has been supposed, that the variolous matter produced its effects by fermentation, and there are some at this day, who believe this doctrine. If it was in any degree true, the quantity of small pox ought to bear some proportion, to the violence of the fever, or first process; and the fermentation ought to produce on the surface, matter at once fully formed; instead of which we observe a regular process is necessary, to maturate the pustules which first appear small and inflamed. The ingenious and candid baron DIMSDALE, has remarked,

§ that

that the future disease, was generally in the inverse ratio of the early symptoms. In proportion, as the symptoms in the arm, where the virus was inserted, were violent and rapid, in their progress, in the same proportion the future disease was mild and secure. This would not happen, if the disease was produced by fermentation. It is curious that matter in a formed state entering the circulation, as it does when the small pox disappears, produces so little danger. It is true, there is some commotion, as the secondary fever, seems evidently the consequence. It is also remarkable, that the secondary fever, is more severe, when the matter absorbed has not undergone the full process of maturation.

Nor can we explain, how maturation diminishes the virulence of the virus, with respect to one individual; and yet possesses its full activity, when again applied to another person, to give the contagion. The kind of small pox is evidently connected with all the circumstances, which regulate inflammation. It is the effect of a peculiar action in the vessels of the skin; and therefore, that action must be modified, when

when we would attempt to change the qualities of the matter produced. All the improvements of Baron DIMSDALE, have been directed to regulate the state of the vessels on the surface; and could in no way affect the variolous matter itself; and I have no doubt, but still greater improvements may be yet made from the use of mercury.

NOTE III.

Among the impediments which retard the progress of medicine, we have not marked *false records*, which most unfortunately fill many of our medical histories. These present, if I may so speak, false facts, or facts so represented as to mislead; and not only perplex the physician, but render his efforts dangerous. It is not possible on any other supposition to account for the total failure of medicines, recorded universally successful, in the hands of some practitioners, and totally useless in the hands of others. The history of cicuta is one example of this kind, and the success of corrosive sublimate, in all cutaneous diseases, another. Men, who thus deliberately mislead, and disguise truth, for the sake of theory or system, are atrocious conspirators against the lives of the human race, and pollute the only true source of knowledge. MEAD, and VAN SWIETEN, STORCK, and many others, have given us records of this kind. We cannot be too minute in describing the effects of medicines which are universally recommended; and all the circumstances of the persons to whom they are prescribed. It has been an unfortunate practice to conceal *unsuccessful cases*; and one side of the question has been only exhibited. This is an idle vanity of success, which is soon detected, whilst the veracity of the practitioners is rendered very questionable. It answers the purposes of empericism, but should be spurned and rejected by a liberal profession.

NOTE IV.

In the enumeration of cases, where a sudden change, not founded on direct indication, produced very remarkable effects, I omitted to mention one of a very singular nature. A soldier in the Welch fusileers, the

the 23d Regiment had been for a long time affected with epileptic fits, which obferved very regular periods in returning. His companions however believed, that he indulged them himfelf, by giving way too much, when he found the fits coming on. He embarked on board a veffel bound from Jamaica to St. Domingo, when the fits obferved their ufual periods. One of the foldiers refolved to play him a trick; he made a poker red hot in the cook's furnace, and whilft the poor fufferer was grafping round in convulfive motion, he put the poker into his hand, which he firmly grafped, leaving on it all the fkin, and a good deal of the flefh of his fingers. This foon waked him from the epilepfy; his hand was cured, and the difeafe never more returned. This cafe was related to me by two refpectable officers now living, who were prefent. The man had been affected for feveral years with this dreadful difeafe. I do not adduce this cafe, as an inftance of practice meriting imitation; but to fhow the force of very powerful and fudden changes.

NOTE V.

IT may perhaps be objected to the mode of prescribing, for altering morbid action and producing changes, that we cannot often say, what kind of action really exists, whether we ought to quicken or to restrain movements. An ingenious writer observes, " that in a machine extremely com-
" plex, formed by the combination of nu-
" merous matter, diversified in their proper-
" ties, in their proportions, in their modes
" of action; the motions necessarily be-
" come extremely complicated, their dul-
" ness as well as their rapidity, frequently
" escape the observation of those themselves
" in whom they take place." Nothing can apply more strictly to the movements of the human body, whether healthy or morbid. There are indeed few instances in which we can perceive or ascertain precisely the kind of action which prevails. Habit, in this, as in many other instances tyrannizes over our native sensibility, and deprives us of that acuteness which accompanied birth. We evidently mark the sensibility of infants, which is affected by the slightest

flighteft change. The impreffion of the air, the expanfion of the lungs, the action of light, the increafed force of the heart, and perhaps the vibration of their arteries, appear in them to create uneafinefs. But habit foon renders us infenfible to their movements. We feel the contractions of the heart, only when its palpitations are unufual. We refpire without confcioufnefs; and walking is often performed, when we are not by any means fenfible of any exertion. Confiderable changes, are thus conftantly going forward, without our being confcious of their exiftence. We cannot employ our fenfes, to afcertain thefe changes, but we are taught by reafon and analogy, that they really exift.

The philofopher who would confine himfelf, to what his fenfes diftinctly unfolded, might indeed be more accurate in his purfuits; but his knowledge would be very limited, without admitting analogy and probability. Without thefe, fcience would be confined within very narrow limits.

But we are fometimes enabled to fay with fome precifion, what kind of action really

really exifts, in particular circumftances of the human body. Thus, in inflammations of the active kind, we evidently fee, and feel, the action of the arteries increafed. But there is fomething attending this action, which we do not comprehend. We cannot afcertain the peculiar modification of it, which induces it to produce pus in thofe inflamed parts. This is the myfterious part of the bufinefs. In fevers, the pulfe is often as frequent, and ftrong as in inflamed tumours; yet pus is not produced, becaufe a fomething which exifted in the pulfe connected with inflammation, does not exift in fevers. In running or dancing, the pulfe is often accelerated, and beats as rapidly for a time as in Fever; yet the head-ach and laffitude, with the other characters of real fever, are abfent; becaufe the peculiar ftate, in which fever confifts, is not at the time prefent. A flow of tears from the eye is caufed by an increafed action in the lachrymal veffels; but acrid fubftances applied to the eye do not produce fuch a plentiful flow, as when the tears have been caufed by grief or forrow. Becaufe, the peculiar action which produced them, does not exift in the application of acrid fubftances.

In uterine hæmorrhage, we can often distinguish two states of action; the one, an increased energy; the other, a diminished action in the vessels. We infer the existence of the first, from a hard rapid pulse, full and tense. We infer the second state, when the pulse is low, flat, and weak; and from the effects of astringents, which in these circumstances usually put a stop to the hæmorrhage. Observation alone can inform us, on these points. But granting, what we must allow, that we seldom know the precise mode of action, we do well when we effect a change, as the system is then more ready to adopt its own healthy and proper movements. Every means, then, ought to be employed, that afford any prospect of effecting these changes; so necessary to the banishment of disease, and the establishment of health.

NOTE VI.

On a careful perusal of Dr. JACKSON's book on Fevers; it affords me the greatest satisfaction to find, a coincidence of thinking, in many particulars which I have treated.

treated. He is the first who boldly pushed cold bathing in fevers, to an extent unknown to former practitioners; he has explained the incessant vomiting and its phænomena in the Yellow Fever on the principles I have been endeavouring to establish. From him I have derived many useful hints, which I had constantly in view, in the course of my experience. Whatever I have been able to observe confirms the general accuracy of his remarks; and I hold this no small proof of the fidelity of what I have related. From Mr. JOHN HUNTER I drew my first notions of morbid action, and endeavoured to apply his doctrine more extensively in diseases; especially in fevers. To follow such leaders, is at least meritorious; and to enlarge or confirm doctrines of which they laid down the elements, may possess ultimately more use than novelty.

FINIS.

Lately Published,
By CADELL and DAVIES, Strand.

A VOYAGE to SAINT DOMINGO,

In the Years 1788, 1789, and 1790:

By *Francis Alexander Stanislaus*, Baron de Wimpffen.

Translated from the original Manuscript, which has never been published; with a Map of the Island.
6 *s.* in Boards.

ALSO,

1. MEDICINA NAUTICA: an Essay on the Diseases of Seamen; comprehending the History of Health in His Majesty's Fleet under the Command of Richard Earl Howe, Admiral. By *Thomas Trotter*, M. D. Member of the Royal Medical Society, an Honorary Member of the Royal Physical Society, &c. Physician to the Fleet. 8vo. 7 *s.* in Boards.

2. Discourses on the Nature and Cure of Wounds. I. Of Generals. Of procuring Adhesion, Wounded Arteries, Gunshot Wounds, Wounds with Sword, &c. The Medical Treatment of Wounds. II. Of Particulars. Of Wounds of the Breasts, Wounds of the Belly, Stitching an Intestine, Wounds of the Head, Wounds of the Throat.
III. Of

III. Of dangerous Wounds of the Limbs, of the Question of Amputation. By *John Bell*, Surgeon. One Volume, Royal 8vo. 7 s. 6 d. in Boards.

⁎ In this Book are contained all those Accidents of Practice and lesser Operations, which do not belong to a System of Surgery, but which, as they occur more frequently, are the more important. It is hoped the Work will be found particularly useful to Country Surgeons, and to young Men entering into the Army and Navy.

3. The Anatomy of the Bones, Muscles, and Joints; being the First Volume of The Anatomy of the Human Body. By *John Bell*, Surgeon. Royal 8vo. 10 s. in Boards.

4. The Anatomy of the Heart and Arteries; being the Second Volume of the same Work. With Plates, 12 s. in Boards.

The following valuable BOOKS are printed for T. CADELL, Jun. and W. DAVIES (Succeſſors to Mr. CADELL) in the Strand, 1797.

HISTORY, VOYAGES, AND TRAVELS.

THE Hiſtory of England, from the Invaſion of Julius Cæſar to the Revolution in 1688. By David Hume, Eſq. A new Edition, printed on fine Paper, with many Corrections and Additions; and a complete Index, 8 vol. Royal Paper, 4to, with fine Impreſſions of the Plates, 10l 10s

⁂ Another Edition, on ſmall Paper, 4l 10s

✠✠✠ Another Edition in 8 vol. 8vo, with the Plates, 2l 16s

The Hiſtory of England, from the Revolution to the Death of George II. forming a Continuation of Mr. Hume's Hiſtory; with Plates. By T. Smollet, M. D. 5 vol. 8vo, 1l 15s

The Hiſtory of Political Tranſactions and of Parties, from the Reſtoration of King Charles II. to the Death of King William. By Thomas Somerville, D. D. 4to, 1l 5s

The Hiſtory of Scotland, during the Reign of Queen Mary and of King James VI. till his Acceſſion to the Crown of England; with a Review of the Scottiſh Hiſtory previous to that Period; and an Appendix, containing Original Papers, 2 vol. 4to. By William Robertſon, D. D. 5th Edition, 1l 10s

⁂ Another Edition in 2 vol. 8vo, 14s

✠✠✠ Another Edition in one volume, 7s 6d

The Hiſtory of the Reign of the Emperor Charles V. with a View of the Progreſs of Society in Europe, from the Subverſion of the Roman Empire to the Beginning of the 16th Century. By W. Robertſon, D. D. Embelliſhed with 4 Plates, elegantly engraved, 3 vol. 4to, 3l 3s

⁂ Another Edition in 4 vol. 8vo, 1l 4s

The Hiſtory of America, vol. I. and II. By W. Robertſon, D. D. Illuſtrated with Maps, 2l 2s

⁂ Another Edition in 3 vol. 8vo, 1l 1s

An Hiſtorical Diſquiſition, concerning the Knowledge which the Antients had of India, and the Progreſs of Trade with that Country, prior to the Diſcovery of the Paſſage to it by the Cape of Good Hope. With an Appendix, containing Obſervations on the Civil Policy—the Laws, and Judicial Proceedings—the Arts—the Sciences, and Religious Institutions of the Indians. By William Robertson, D. D. Illustrated with Maps, 4to, 18s

⁎ Another Edition in one vol. 8vo, 8s

The History of Greece. By William Mitford, Esq. Vol. I. and II. 4to, 1l 19s in boards.

⁎ The two Volumes include the History of Greece from the earliest Accounts to the End of the Peloponnesian War; and it is intended to continue the History till the Reduction of Achaia into a Province of the Roman Empire.

The same Work in 4 vol. 8vo, 1l 8s

⁎ The third Volume in 4to, is now in the Press.

The History of Ancient Greece, its Colonies and Conquests; from the earliest Accounts till the Division of the Macedonian Empire in the East: including the History of Literature, Philosophy, and the Fine Arts. By John Gillies, LL D. F. R. S. with a Head of the Author, and Maps adapted to the Work, 4 vol. 3d Edit. 1l 8s

A View of the Reign of Frederic II. of Prussia, with a Parallel between that Prince and Philip II. of Macedon. By J. Gillies, LL. D. F. R. S and S. A. 8vo, 7s

The History of the Reign of Philip II. King of Spain. By Robert Watson, LL. D. Professor of Philosophy and Rhetoric at the University of St. Andrew, 2 Edition, 2 vol. 4to, 2l 2s—Another Edition in 3 vol. 8vo, 18s

The History of the Decline and Fall of the Roman Empire. By Edward Gibbon, Esq. 6 vol. which complete a Period of History from the Age of Trajan and the Antonines, to the taking of Constantinople by the Turks, and the Establishment at Rome of the Dominion of the Popes. Adorned with a Head of the Author, and Maps adapted to the Work, 6l 6s in Boards.

⁎ The 4th, 5th, and 6th Vol. may be had separate, to complete Sets, 3l 3s in Boards.

†‡‡ Another Edition, complete in 12 vol. 8vo, 3l 12s

Also an Abridgment of this Work for the Use of young Persons, 2 vol. 8vo, 14s

The History of Rome, from the Foundation of the City by Romulus, to the Death of Marcus Antoninus, 3 vol. 8vo, 1l 1s. This Work, with the Abridgment of Mr. Gibbon's History, forms a complete Roman History, in 5 vol. 8vo.

The Chronology and History of the World, from the Creation to the Year of Christ 1790, illustrated on 56 Copperplate Tables, with 16 Maps of Ancient and Modern Geography. By the Reverend John Blair, LL. D. 3l 13s 6d

The History of France from the Accession of Henry III. to the Death of Louis XIV. preceded by a View of the Civil, Military, and Political State of Europe, between the Middle and the Close of the Sixteenth Century. By Nathaniel William Wraxall. Vol. I. II. and III. 4to, 3l 3s in Boards.

Memoirs of Great Britain and Ireland, from the Dissolution of the last Parliament of Charles II. until the Capture of the French and Spanish Fleets at Vigo. By Sir John Dalrymple, Bart. 3d Edition, with Appendixes complete, 3 vol. 1l 16s

The History of England, from the Earliest Accounts of Time to the Death of George II. adorned with Heads, elegantly engraved. By Dr. Goldsmith, 4 vol. 1l 4s

An Abridgment of the above Book, by Dr. Goldsmith, adorned with Cuts, for the Use of Schools, 3s 6d

An Ecclesiastical History, Ancient and Modern, from the Birth of Christ to the Beginning of the Present Century. In which the Rise, Progress, and Variations of Church Power are considered, in their Connexion with the State of Learning and Philosophy, and the Political History of Europe during that Period. By the late learned John Lawrence Mosheim, D. D. Translated, and accompanied with Notes and Chronological Tables, by Archibald Maclaine, D. D. A new Edition, corrected and improved, 6 vol. 2l 2s

A Summary of Geography and History, both Ancient and Modern; containing an Account of the most illustrious Nations in Ancient and Modern Times; their Manners and Customs; the local Situation of Cities, especially of such as have been distinguished by memorable Events. With an Abridgment of the Fabulous History or Mythology of the Greeks, &c. The Whole chiefly designed to connect the Study of Classical Learning with that of General Knowledge. By Alexander Adam, LL. D. Rector of the High School of Edinburgh. Illustrated with a new Set of Maps, 9s

Roman Antiquities; or an Account of the Manners and Customs of the Romans; respecting their Government, Magistracy, Laws, Religion, Games, Military and Naval Affairs, &c. &c. Designed chiefly to illustrate the Latin Classics. By Alexander Adam, LL. D. 3d Edition, 7s 6d

4   BOOKS PRINTED FOR CADELL AND DAVIES.

Bibliotheca Claffica; or a Claffical Dictionary, containing a full Account of all the proper Names mentioned in ancient Authors. To which are fubjoined, Tables of Coins, Weights, and Meafures, in Ufe among the Greeks and Romans. By J. Lempriere, A. M. of Pembroke-College, Oxford. 2d Edition, 9s

A Philofophical and Political Hiftory of the Settlements and Trade of the Europeans in the Eaft and Weft Indies. Tranflated from the French of the Abbé Reynal, by J. Juftamond, M. A. A new Edition, carefully revifed, in 8 vol. 8vo, and illuftrated with Maps, 2l 8s

Sketches of the Hiftory of Man, by the Author of the Elements of Criticifm. 3d Edition, 4 vol. 1l 8s

An Account of the Voyages undertaken by Order of his prefent Majefty, for making Difcoveries in the Southern Hemifphere, and fucceffively performed by Commodore Byron, Captain Wallis, and Captain Carteret, in the Dolphin, the Swallow, and Endeavour. Drawn up from the Journals which were kept by the feveral Commanders, and from the Papers of Jofeph Banks, Efq. and Dr. Solander. By John Hawkefworth, LL. D. Illuftrated with Cuts, and a great Variety of Charts and Maps (in all 52 Plates) 3 vol. 4to, Price 3l 12s

An Account of a Voyage towards the South Pole and round the World, performed in his Majefty's Ships the Refolution and Adventure, in the Years 1772, 1773, 1774, and 1775. Written by James Cook, Commander of the Refolution. In which is included, Captain Furneaux's Narrative of his Proceedings in the Adventure during the Separation of the Ships. 2 vol. Royal 4to, with Maps and Charts, Portraits of Perfons, and Views of Places, drawn during the Voyage, by Mr. Hodges, and engraved by the moft eminent Mafters, 2l 12s

A Journey from Prince of Wales's Fort, in Hudfon's Bay, to the Northern Ocean. Undertaken by Order of the Hudfon's Bay Company, for the Difcovery of Copper-Mines: a North-Weft Paffage, &c. in the Years 1769, 1770, 1771, and 1772. By Samuel Hearne, 4to, with Charts and other Plates, 1l 11s 6d

The Environs of London: being an Hiftorical Account of the Towns, Villages, and Hamlets, within 12 Miles of that Capital; interfperfed with Biographical Anecdotes. By the Rev. D. Lyfons, A. M. F. R. S. 3 vol. 4to, 4l 14s 6d in Boards.

Travels in Portugal, through the Provinces of Entre Douro e Minho, Beira, Eftremadura, and Elem-tejo, in the Years 1789 and 1790: confisting of Obfervations on the Manners, Cuftoms, Trade, Public Buildings, Arts, Antiquities, &c. of that Kingdom. With 24 Plates of Views, &c. By James Murphy, Architect, 4to, 1l 11s 6d

Travels through Spain, in the Years 1775 and 1776. In which feveral Monuments of Roman and Moorifh Architecture are illuftrated by accurate Drawings taken on the Spot. By Henry Swinburn, Efq. 2d Edition, 2 vol. 14s

Travels in the Two Sicilies. By Henry Swinburn, Efq. in the Years 1777, 1778, 1779, and 1780. With a Map of the Two Sicilies, and 22 Plates of Views, &c. 2d Edition, 4 vol. 1l 8s

Travels through various Provinces of the Kingdom of Naples, in 1789. By Charles Ulyffes, of Salis Marfchlins. Tranflated from the German, by Anthony Aufrere, Efq. and illuftrated with Plates, 8vo, 9s

⁎ This Volume forms a proper Supplement to Mr. Swinburn's Travels in the Two Sicilies.

Travels into different Parts of Europe, in the Years 1791 and 1792. With Familiar Remarks on Places, Men, and Manners. By John Owen, A. M. late Fellow of Corpus Chrifti College, Cambridge. 2 vol. 16s

Travels in Switzerland, in a Series of Letters to William Melmoth, Efq. from William Coxe, M. A. F. R. S. F. A. S. Rector of Bemerton, &c. 3 vol. with a large Map of Switzerland, and other Plates, 1l 7s

Travels into Poland, Ruffia, Sweden, and Denmark, interfperfed with Hiftorical Relations and Political Enquiries, illuftrated with Maps and Engravings. By William Coxe, A. M. F. R S. &c. 5 vol. 3d Edit. 1l 17s 6d

A new Volume, being the 3d in 4to, and the 5th in 8vo, of the above Work, with a Map of Southern Norway, and other Plates, may be had feparate.

An Account of the Ruffian Difcoveries between Afia and America; to which are added, the Conqueft of Siberia, and the Hiftory of the Tranfactions and Commerce between Ruffia and China. By William Coxe, A. M. F. R. S. Illuftrated with Charts, and a View of a Chinefe Town. 3d Edit. 7s 6d

A complete Tranflation of the Count de Buffon's Natural Hiftory, from the 4th Edit. in 16 vol. 4to, with occafional Notes and Obfervations. By William Smellie, Member of the Philofophical and Antiquarian Societies of Edinburgh. Illuftrated with 300 Copperplates, 9 vol, 4l 1s

A new Translation of the Count de Buffon's Natural History of Birds. Illustrated with near 300 Engravings, and a Preface, Notes, and Additions by the Translator, 9 vol. 4l 1s

A new System of the Natural History of Quadrupeds, Birds. Fishes, and Insects. With about 150 Copperplates, 3 large volumes 8vo, 1l 16s

A Tour through Sicily and Malta, in a Series of Letters to William Beckford, Esq. from P. Brydone, F. R. S. 2 vol. illustrated with a Map. 3d. Edition, 12s

Observations and Reflections made in the Course of a Journey through France, Italy, and Germany. By Heister Lynch Piozzi, 2 vol. 14s

Observations made in a Tour from Bengal to Persia, in the Year 1886-7; with a short Account of the Remains of the celebrated Palace of Persepolis, and other interesting Events. By William Francklin, Ensign on the Hon. Company's Bengal Establishment, lately returned from Persia. 2d Edit. 8vo, 7s

A View of Society and Manners in France, Switzerland, and Germany, with Anecdotes relating to some eminent Characters. By John Moore, M. D. 2 vol. 7th Edit. 12s

A View of Society and Manners in Italy, with Anecdotes relating to some eminent Characters. By John Moore, M. D. 2 vol. 4th Edition, 14s

A Journey to the Western Isles of Scotland. By the Author of the Rambler, 6s

The State of the Prisons in England and Wales, with Preliminary Observations, and an Account of some Foreign Prisons and Hospitals. By John Howard, F. R. S. 4to, 4th Edit. with all the Plates complete, 1l 5s

An Account of the principal Lazarettos in Europe; with various Papers relative to the Plague, &c. by the same. 2d Edit. with all the Plates complete, 1l 5s

DIVINITY.

Isaiah: a new Translation, with a Preliminary Dissertation, and Notes Critical, Philological, and Explanatory. By Robert Lowth, D. D. F. R. S. Londin. & Goetting. late Lord Bishop of London. 3d Edit. 2 vol. 8vo, 14s

Jeremiah and Lamentations: a new Translation, with Notes Critical, Philological, and Explanatory. By Benjamin Blayney, D. D. &c. &c. 4o, 1l 5s

The Four Gospels, tranflated from the Greek; with Preliminary Diſſertations, and Notes Critical and Explanatory. By George Campbell, D. D. F. R. S. Principal of Mariſchal College, Aberdeen, 2 vol. 4to, 2l 10s

Sermons on ſeveral Subjects. By the Right Rev. Beilby Porteus, D. D. Biſhop of London. 6th Edit. 2 vol. 13s

⁎ The ſecond Vol. may be had ſeparate, 6s in Bds.

An Introduction to the Study of the Prophecies concerning the Chriſtian Church, and in particular concerning the Church of Papal Rome. By Richard Hurd, D. D. now Lord Biſhop of Worceſter. 3d Edit. 2 vol. 7s

Sermons preached at Lincoln's Inn Chapel, between the Years 1765 and 1776. By Richard Hurd, D. D. Lord Biſhop of Worceſter. 2 Edit. 3 vol. 18s

The Works of the Right Rev. Jonathan Shipley, D. D. Lord Biſhop of St. Aſaph, 2 v. 8vo, with a Portrait, 12s

Sermons by Hugh Blair, D. D. one of the Miniſters of the High Church, and Profeſſor of Rhetoric and Belles Lettres in the Univerſity of Edinburgh. 19th Edition, 4 vol. 1l 8s

⁎ The 4th Volume may be had ſeparate, 6s in Boards.

Sermons by William Leechman, D. D. with ſome Account of the Author's Life, and of his Lectures. By James Wodrow, D. D. 2 vol. 14s

Sermons by the late Rev. John Dryſdale, D. D. F. R. S. Ed. one of the Miniſters of Edinburgh, &c. &c. With an Account of his Life and Character, by Andrew Dalzel, M. A. F. R. S. Edinburgh, Profeſſor of Greek, &c. in the Univerſity of Edinburgh, 2 vol. 14s

Sermons by George Hill, D. D. F. R. S. Edinburgh, Principal of St. Mary's College in the Univerſity of St. Andrew; one of the Miniſters of that City, and one of his Majeſty's Chaplains in Ordinary for Scotland, 7s

Diſcourſes, chiefly on the Evidences of Natural and Revealed Religion. By John Sturges, LL. D. 7s

Diſcourſes on various Subjects. By Jacob Duché, M. A. 3d Edit. 2 vol. 14s

Sermons on different Subjects, left for Publication by John Taylor, LL. D. late Prebendary of Weſtminſter, Rector of Boſworth, Leiceſterſhire, and Miniſter of St. Margaret, Weſtminſter. Publiſhed by the Reverend Samuel Hayes, A. M. 2 vol. 12s

⁎ Theſe Volumes include the Sermon written by Dr. Johnſon, for the Funeral of his Wife; and

BOOKS PRINTED FOR CADELL AND DAVIES.

all the Sermons exhibit strong internal evidence of their having been carefully revised, at least, if not wholly written by that eminent Moralist, who had been, for a great Number of Years, in Habits of close Intimacy with the Divine whose Name they bear.

Sermons on the Present State of Religion in this Country, and on other Subjects. By the Rev. Septimus Hodson, M. B. Rector of Thrapston, Chaplain to the Asylum, and Chaplain in Ordinary to his Royal Highness the Prince of Wales, 8vo, 5s

One hundred Sermons on Practicable Subjects, extracted chiefly from the Works of the Divines of the last Century. By Dr. Burn, 4 vol. 1l 4s

Sermons by the late Lawrence Sterne, M. A. 6 vol. 18s

Sermons on the Christian Doctrine, as received by the different Denominations of Christians, &c. &c. By R. Price, D. D. LL. D. F. R. S. &c. 2 Edit. 6s

Sermons on various Subjects, and preached on several Occasions. By the late Rev. Thomas Francklin, D. D. 4th Edition, 3 vol. 1l 4s

Sermons on the Relative Duties. By the same, 6s

Sermons to Young Men. By W. Dodd, LL. D. Prebendary of Brecon, and Chaplain in Ordinary to his Majesty. 3d Edit. 7s 6d

Four Dissertations. I. On Providence. II. On Prayer. III. On the Reasons for expecting that virtuous Men shall meet after Death in a State of Happiness. IV. On the Importance of Christianity, the Nature of Historical Evidence, and Miracles. By Richard Price, D. D. F. R. S. 4th Edit. 8vo, 6s

Sermons to Young Women. By James Fordyce, D. D. 2 vol. 6th Edit. 7s

Addresses to Young Men, by the same. 2 vol. 8s

Sermons on Various Subjects. By the late John Farquhar, A. M. Carefully corrected from the Author's MSS. by George Campbell, D. D. and Alexander Gerrard, D. D. 4th Edit. 7s

A Review of the Principal Questions in Morals. By Richard Price, D. D. F. R. S. 3 Edit. corrected, 7s

MISCELLANIES, BOOKS OF ENTERTAINMENT, &c.

The Works of the late Right Hon. Henry St. John, Lord Viscount Bolingbroke; containing all his Political and Philosophical Works: a new and elegant Edit. 5 vol. 4to.

⁎ Another Edition in 11 vol. 8vo, 2l 16s

The Works of Francis Bacon, Baron of Verulam, Vifcount St. Alban's, and Lord High Chancellor of England. 5 vol. Royal Paper, 4to.

An Inquiry into the Nature and Causes of the Wealth of Nations. By Adam Smith, LL. D. F. R. S. 3 vol. 6th Editon, 8vo, 1l 1s

An Inquiry into the Principles of Political Œconomy; being an Essay on the Science of Domestic Policy in Free Nations; in which are particularly considered, Population, Agriculture, Trade, Industry, Money, Coin, Interest, Circulation, Banks, Exchange, Public Credit, Taxes, &c. By Sir James Stuart, Bart. 2 vol. Royal Paper. 4to, 2l 2s in Boards.

Essays and Treatises on several Subjects. By David Hume, Esq. with his last Corrections and Additions. 2 vol. 4to, 1l 16s

⁎ Another Addition, in 2 vol. 8vo, 14s

Moral and Political Dialogues, with Letters on Chivalry and Romance. By Richard Hurd, D. D. now Lord Bishop of Worcester. 3 vol. 10s 6d

An Essay on the History of Civil Society. By Adam Ferguson, LL. D. 4th Edit. 7s

Zeluco. Various Views of Human Nature, taken from Life and Manners, foreign and domestic. 2d Edit. 2 vol. 14s

The Theory of Moral Sentiments. By Adam Smith, LL. D. F. R. S. 5th Edit. 2 vol. 14s

The Elements of Moral Science. By James Beattie, LL. D. Professor of Moral Philosophy and Logic in Marischal College, Aberdeen. 2 vol. 8vo, 15s

The Works of Alexander, Pope, Esq. with his last Corrections, Additions, and Improvements, as they were delivered to the Editor a little before his Death; together with the Commentary and Notes of Dr. Warburton. With Cuts. In 9 large Volumes, 8vo, 2l 14s

The same in 6 Volumes, 12mo, 18s

A complete and elegant Edition of the English Poets, printed in 75 Pocket Volumes, on a fine Writing Paper. Illustrated with Heads, engraved by Bartolozzi, Caldwall, Hall, Sherwin, &c. &c. with a Preface Biographical and Critical to each Author. By Samuel Johnson, LL. D. 13l 2s 6d

Miscellaneous Works of Edward Gibbon, Esq. With

Memoirs of his Life and Writings, composed by himself; illustrated from his Letters, with Occasional Notes and Narrative. By John Lord Sheffield. 2 vol. 4to, 2l 10s in Boards.

The Works of Soame Jenyns, Esq. including several Pieces never before published. To which are prefixed, short Sketches of the History of the Author's Family, and also of his Life. By Charles Nalson Cole, Esq. with a Head of the Author. 4 vol. 2d Edition, 1l

Letters to and from the late Samuel Johnson, LL D. To which are added, some Poems never before printed. Published from the Original in her Possession. By Hester Lynch Piozzi. 2 vol. 4s

The Lives of the most eminent English Poets; with Critical Observations on their Works. By Samuel Johnson, LL. D. 4 vol. new Edition, 1l 4s

The Life of Milton, in three Parts. To which are added Conjectures on the Origin of Paradise Lost, with an Appendix. By William Hayley, Esq. 4to, 18s

Anecdotes of some distinguished Persons, chiefly of the present and two preceding Centuries. Adorned with Sculptures. 4 vol. 3d Edit. 1l 8s in Boards.

*** The fourth Volume may be had separate.

The Works of Mr. Thomson, complete, elegantly printed on a fine Writing Paper, with Plates, and a Life of the Author. 3 vol. 1l 4s

Another Edition, 3 vol. Crown, 18s; or 2 v. common 7s

The Seasons, in a Twelves Edit. Price only 2s 6d

Another Edition of the Seasons, in a smaller Size, printed on a fine Paper, with new Plates, 6s

The Triumphs of Temper: a Poem in 6 Cantos. By William Hayley, Esq. 6th Edit. 7s 6d

Elegiac Sonnets. By Charlotte Smith. 5th Edit. with additional Sonnets, and other Poems, 7s 6d

The Pleasures of Memory, with some other Poems. By S. Rogers, Esq. 7th Edit. 7s 6d

The Pleasures of Imagination. By Mark Akenside, M. D. To which is prefixed, a Critical Essay on the Poem. By Mrs. Barbauld, 7s 6d

The Art of preserving Health. By John Armstrong, M. D. With a Critical Essay on the Poem, by J. Aikin, M. D. 7s 6d

The Spleen, and other Poems of Matthew Green. With a Critical Essay by Dr. Aikin, 6s

The Shipwreck, a Poem, in 3 Cantos. By a Sailor, 6s

*** The seven last-mentioned Works are printed in a most beautiful and uniform Manner, and are all embellished with very fine Plates.

Essays on various Subjects, principally designed for young Ladies. By Hannah More, 3s sewed, 2d Edit.

Adelaide and Theodore, or Letters on Education: containing all the Principles relative to the different Plans of Education. Translated from the French of Madame la Comtesse de Genlis. 3 vol. 3 Edit. 10s 6d

The Moral Miscellany, or a Collection of Select Pieces, in Prose and Verse, for the Instruction and Entertainment of Youth. 3d Edit. 3s

An Historical Miscellany. 3d Edit. 3s

The Poetical Miscellany; consisting of Select Pieces from the Works of the following Poets, viz. Milton, Dryden, Pope, Addison, Gay, &c. 2 Edit. 3s

A Father's Legacy to his Daughters, by the late Dr. Gregory, of Edinburgh. With a Frontispiece, 2s sewed

The Mirror: a Periodical Paper, published at Edinburgh in the Years 1779 and 1780. 3 vol. 8th Edit. 9s; or 2 vol. large 8vo, 14s

The Lounger: a Periodical Paper. By the Authors of the Mirror. 4th Edit. 10s 6d; or 2 vol. large 8vo, 14s

The Adventurer, by Dr. Hawkesworth, 4 vol. A new Edit. adorned with Frontispieces, 12s; or 3 vol. large 8vo, 1l 1s

The Rambler, 4 vol. with Frontispieces, and a Head of the Author, 12s; or 3 vol. large 8vo, 1l 1s

A complete Edition of the Works of Lawrence Sterne, M. A. containing his Tristram Shandy, Sentimental Journey, Letters, &c. &c. Adorned with Plates, designed by Hogarth, Rooker, Edwards, &c. 10 vol. 2l

The Man of Feeling, a Novel. A new Edit. 3s

The Man of the World, by the Author of the Man of Feeling, 2 vol. 6s

Julia de Roubigne, by the same, 2 vol. 6s

Sentimental Journey, 2 vol. A new Edit. with Frontispieces, 5s—Another Edit. 3s

Tristram Shandy, 6 vol. 18s

The Recess; or, a Tale of other Times. By the Author of the Chapter of Accidents. 4th Edit. 3 vol. 13s 6d

Julia, a Novel, interspersed with some Poetical Pieces. By Helen Maria Williams, 2 vol. 7s

Emmeline, the Orphan of the Castle. By Charlotte Smith, 4 vol. 3d Edit. 14s

12   BOOKS PRINTED FOR CADELL AND DAVIES.

Ethelinde; or, the Recluse of the Lake, by the same. 5 vol. 2d Edit. 17s 6d

Celestina, a Novel, by the same, 4 vol.   2d Edit. 14s

The Banished Man, a Novel, by the same. 4 vol. 2d Edit. 16s

Extracts, elegant, instructive, and entertaining, in Prose, from the most approved Authors; disposed under proper Heads, with a View to the Improvement and Amusement of young Persons, one vol. Royal 8vo, 14s

Extracts in Poetry, upon the same Plan, 16s

Epistles, elegant, familiar, and instructive, selected from the best Writers, ancient and modern: a proper Companion to the two preceding Works, 9s

LAW.

Commentaries on the Laws of England, in Four Books. By Sir William Blackstone, Knight, one of the Justices of his Majesty's Court of Common Pleas. Twelfth Edition, with the Author's last Corrections, and Notes and Additions by Edward Christian, Esq. Barrister at Law, and Professor of the Laws of England in the University of Cambridge. 4 large Volumes, 8vo, with 14 Portraits, elegantly engraved, 1l 16s

Tracts, chiefly relating to the Antiquities of the Laws of England. By Judge Blackstone, 4to, 1l 1s

A Digest of the Laws of England. By the Right Hon. Sir John Comyns, Knt. late Lord Chief Baron of his Majesty's Court of Exchequer. 3d Edit. considerably enlarged, and continued down to the present Time, by Stewart Kyd, Barrister at Law, of the Middle Temple, 6 v. Royal 8vo, 4l 4s

Cases in Crown Law, determined by the Twelve Judges, by the Court of King's Bench, and by Commissioners of Oyer and Terminer, and General Gaol Delivery: from the 4th Year of George II. to the 32d Year of George III. By Thomas Leach, Esq. of the Middle Temple, Barrister at Law. 2d Edit. with Additions, 12s

Cases argued and determined in the High Court of Chancery, in the Time of Lord Chancellor Hardwicke, from the Years 1746-7 to 1755, with Tables, Notes, and References. By Francis Vezey, Esq. 2 vol. 3d Edit. 1l 1s

The Attorney's Vade Mecum, and Client's Instructor, treating of Actions (such as are now most in Use); of prosecuting and defending them; of the Pleadings and

Law, with a Volume of Precedents. By John Morgan, of the Inner Temple, Barrister at Law, 3 vol. 1l 2s

The Justice of Peace; or, Complete Parish Officer. By Richard Burn, LL. D. A new Edition, 4 vol. 1l 12s

Ecclesiastical Law, by the same Author, 4 vol. 1l 8s

A new Law Dictionary, intended for General Use, as well as for Gentlemen of the Profession. By Richard Burn, LL. D. and continued to the present Time, by his Son, 2 vol. 16s

A Digest of the Law of Actions at Nisi Prius. By Isaac Espinasse, Esq. of Gray's Inn, Barrister at Law. 2d Edition, 2 vol. 1l 1s

The Solicitor's Guide to the Practice of the Office of Pleas in his Majesty's Court of Exchequer at Westminster; in which are introduced, Bills of Costs in various Cases, and a Variety of useful Precedents. With a complete Index to the whole. By Richard Edmunds, one of the Attornies of the said Office, 7s

PHYSIC.

Domestic Medicine; or, a Treatise on the Prevention and Cure of Diseases, by Regimen and Simple Medicine. By Wm. Buchan, M. D. of the Royal College of Physicians, Edinburgh. A new Edition, 7s 6d

⁎ This Treatise comprehends not only the Acute, but also the Chronic Diseases; and both are treated at much greater Length than in any Performance of the like Nature. It likewise contains an Essay on the Nursing and Management of Children; with Rules for preserving Health, suited to the different Situations and Occupations of Mankind: and Directions for the Cure of Wounds, the Reduction of Fractures, Dislocations, &c.

Medical Histories and Reflections. By John Ferriar, M. D. Physician to the Manchester Infirmary and Lunatic Hospital, 2 vol. 11s

First Lines of the Theory and Practice of Philosophical Chemistry. By John Berkenhout, M. D. 8vo, with Plates, 7s 6d

The Seats and Causes of Diseases investigated by Anatomy, in five Books; containing a great Variety of Dissections with Remarks. Translated from the Latin of John Baptist Morgagni, Chief Professor of Anatomy, and Prefident of the Univerfity of Padua   By Benjamin Alexander, M. D. 3 vol. 4to, 1l 16s

A full and plain Account of the Gout; from whence will be clearly feen the Folly, or the Bafenefs, of all Pretenders to the Cure of it: in which every Thing material by the beft Writers on that Subject, is taken notice of, and accompanied with fome new and important Inftructions for its Relief; which the Author's Experience in the Gout, above thirty Years, has induced him to impart. By Ferdinando Warner, LL. D. 3d Edit. 5s

A Treatife upon Gravel and on Gout, in which their Sources and Connexion are afcertained; with an Examination of Dr. Auftin's Theory of Stone, and other Critical Remarks. A Differtation on the Bile, and its Concretions; and an Enquiry into the Operations of Solvents. By Murray Forbes, Member of the Surgeons Company, 6s

An Account of the Efficacy of the Aqua Mephitica Alkalina; or, Solution of fixed Alkaline Salt, faturated with fixable Air, in Calculous Diforders, and other Complaints of the Urinary Paffages. By W. Falconer, M. D. F. R. S. Phyfician to the General Hofpital at Bath. 4th Edit. 3s

An Enquiry into the Nature, Caufes, and Method of Cure of Nervous Diforders. By Alexander Thompfon, M. D. 2s

A new Enquiry into the Caufes, Symptoms, and Cure of Putrid and Inflammatory Fevers, &c. &c. By Sir William Fordyce, M. D. 4s

Difcourfes on the Nature and Cure of Wounds. I. Of Generals: of procuring Adhefion, Wounded Arteries, Gun-fhot Wounds, Wounds with Sword, &c. the Medical Treatment of Wounds. II. Of Particulars: of Wounds of the Breaft, Wounds of the Belly, ftitching an Inteftine, Wounds of the Head, Wounds of the Throat. III. Of dangerous Wounds of the Limbs. Of the Queftion of Amputation. By John Bell, Surgeon, 1 vol. Royal 8vo, 7s 6d in Boards.

₊ In this Book are contained all thofe Accidents of Practice and leffer Operations which do not belong to a Syftem of Surgery, but which, as they occur more frequently, are the more important. — This Book, it is hoped, will be found particularly ufeful to Country Surgeons, and to Young Men entering into the Army and Navy.

Engravings explaining the Anatomy of the Bones, Mufcles, and Joints, with copious Descriptions. By John Bell, Surgeon, 4to, 1l 1s in Boards

A Collection of Cases and Observations in Midwifery. By Wm. Smellie, M. D. 3 vol. with Cuts, 1l 1s

PHILOSOPHY, MATHEMATICS, MECHANICS, &c. &c.

Elements of the Philosophy of the Human Mind. By Dugald Stewart, F. R. S. Edin. Professor of Moral Philosophy in the University of Edinburgh. 4to, 1l 5s

Astronomy explained upon Sir Isaac Newton's Principles, and made easy to those who have not studied the Mathematics. By James Ferguson, F. R. S. Illustrated with 28 Copperplates. A new Edit. 8vo, 9s

An easy Introduction to Astronomy, for young Gentlemen and Ladies, by the same. 3d Edit. 5s

An Introduction to Electricity, in 6 Sections, by the same. Illustrated with Plates, 4s

Lectures on Select Subjects in Mechanics, Hydrostatics, Pneumatics, and Optics, with the Use of the Globes, the Art of Dialling, and the Calculation of the Mean Times of New and Full Moons and Eclipses, by the same, 7s 6d

Select Mechanical Exercises, shewing how to construct different Clocks, Orreries, and Sun-Dials, on plain and easy Principles, &c. &c. By the same. With Copperplates, and a short Account of the Life of the Author, 5s

Observations on Reversionary Payments; on Schemes for granting Annuities to Widows, and to Persons in old Age; on the Method of finding the Value of Assurances on Lives and Survivorship; and on the National Debt. To which are added, new Tables of the Probabilities of Life; and Essays on the different Rates of Human Mortality in different Situations, &c. &c. By Richard Price, D. D. F. R. S. A new Edit. 2 vol. 8vo, 15s

The Doctrine of Annuities and Assurances on Lives and Survivorship, stated and explained. By William Morgan, Actuary to the Society for Equitable Assurances on Lives and Survivorships, 8vo, 6s

AGRICULTURE, BOTANY, GARDENING, &c.

The Complete Farmer; or, a General Dictionary of Husbandry in all its Branches; containing the various Methods of cultivating and improving every Species of Land according to the Precepts of both the old and new Husbandry; comprising every Thing valuable in the best Writers on the Subject: together with a great Variety of new Discoveries and Improvements. 4th Edit. considerably enlarged, and greatly improved. By a Society of Gentlemen, Members of the Society for the Encouragement of Arts, Manufactures, and Commerce. With a great Number of Plates, Folio, 2l 2s

Practical Essays on Agriculture: containing an Account of Soils, and the Manner of correcting them; an Account of the Culture of all Field Plants, including the Artificial Grasses, according to the old and new Modes of Husbandry, with every Improvement down to the present Period; also an Account of the Culture and Management of Grass Lands; together with Observations on Inclosures, Fences, Farms, Farm-houses, &c. Carefully collected and digested from the most eminent Authors, with experimental Remarks. By James Adam, Esq. 2 v. 14s

Every Man his own Gardener; being a new and much more complete Gardener's Calendar than any hitherto published; containing not only an Account of what Work is necessary to be done in the Hot-House, Green-House, Shrubbery, Kitchen, Flower, and Fruit-Gardens, for every Month in the Year, but also ample Directions for performing the said Work according to the newest and most approved Methods now in Practice amongst the best Gardeners. By Thomas Mawe, Gardener to his Grace the Duke of Leeds; and other Gardeners. 6th Edit. 5s

An Introduction to Botany; containing an Explanation of the Theory of that Science, and an Interpretation of its Technical Terms; extracted from the Works of Dr. Linnæus, and calculated to assist such as may be desirous of studying the Author's Method and Improvements. With Plates. 3d Edit. with a Glossary, and other Additions. By James Lee, 7s 6d

Synopsis of the Natural History of Great Britain and Ireland: containing a Systematic Arrangement and concise Description of all the Animals, Vegetables, and Fossils, which have hitherto been discovered in these Kingdoms. By John Berkenhout, M. D. 2d Edit. 2 vol. 12s

Clavis Anglica Linguæ Botanicæ; or, a Botanical Lexicon. In which the Terms of Botany, particularly those occurring in the Works of Linnæus, and other modern Writers, are applied, derived, explained, contrasted, and exemplified. by John Berkenhout, M. D. 2d Edit. 6s

Historical and Biographical Sketches of the Progress of Botany in England, from its Origin to the Introduction of the Linnæan System. By Richard Pulteney, M. D. F. R. S. 2 vol. 12s

www.ingramcontent.com/pod-product-compliance
Lightning Source LLC
Chambersburg PA
CBHW030423300426
44112CB00009B/828